Alfred Wilks Drayson

Untrodden Ground in Astronomy and Geology

Giving Further Details of the Second Rotation of the Earth and of the...

Alfred Wilks Drayson

Untrodden Ground in Astronomy and Geology
Giving Further Details of the Second Rotation of the Earth and of the...

ISBN/EAN: 9783337002473

Printed in Europe, USA, Canada, Australia, Japan

Cover: Foto ©berggeist007 / pixelio.de

More available books at **www.hansebooks.com**

UNTRODDEN GROUND

IN

ASTRONOMY AND GEOLOGY

GIVING FURTHER DETAILS OF

THE SECOND ROTATION OF THE EARTH

AND OF THE

IMPORTANT CALCULATIONS WHICH CAN BE MADE BY AID
OF A KNOWLEDGE THEREOF

BY

MAJOR-GENERAL A. W. DRAYSON, F.R.A.S.
LATE R.A.

AUTHOR OF "PRACTICAL MILITARY SURVEYING"
"THE LAST GLACIAL EPOCH," "COMMON SIGHTS IN THE HEAVENS"
"THIRTY THOUSAND YEARS OF THE EARTH'S PAST HISTORY," ETC.

LONDON
KEGAN PAUL, TRENCH, TRÜBNER & CO., Lt?
1890

PREFACE.

IN the following pages some further details will be given of the second rotation of the earth, and the important facts that are revealed by a knowledge thereof.

The reader who wishes to become conversant with this hitherto unknown movement, is recommended to investigate this second rotation by itself; to omit, for the time being, all notice of the earth's daily rotation, and of its annual revolution round the sun, and to examine the details of a slow second rotation of a sphere. It will then be manifest that, whilst the poles of rapid rotation are carried over small arcs by the second rotation, every other point on the surface of the sphere will be carried over arcs by the same movement; but these arcs will differ not only in extent, but in direction, according to their positions relative to the poles of the *second* axis of rotation.

Thus every zenith of every locality on earth is differently affected during the year by the second rotation. The zenith of the north pole of daily rotation is carried annually over an arc of about 20″, which arc is traced nearly down a meridian of twenty-four hours right ascension; but a zenith 90° from the pole of *second* rotation is carried over an arc of about $40\tfrac{9}{10}''$, the direction of the arcs being nearly in opposition to the daily rotation.

Two localities on the same meridian of terrestrial longitude, but differing in terrestrial latitude, may have their zeniths affected by the second rotation in a very different manner. One zenith may be carried directly north; the other zenith may be carried obliquely towards the east or west. These varied results will be most marked on those meridians which have about fifteen, sixteen, seventeen, nineteen, twenty, and twenty-one hours of right ascension. The detailed effect will be found fully explained in Chapter XIII.

By a knowledge of this second rotation results can be arrived at with minute accuracy by calculation, which results have hitherto been obtained, and then only for very short periods, by continued and laborious observation.

The effect of the second rotation as at present occurring, when traced back during 20,000 years, is to have produced a very great annual variation of climate in both hemispheres of the earth, more particularly in latitudes from 50° up to the poles, an arctic climate in winter reaching to about 54° latitude in each hemisphere, and these annual variations prevailing during about 15,000 years, the date of the height of this period being about 13,544 B.C.

Whilst, therefore, a knowledge of this movement of the earth enables a geometrician to arrive by calculation at results hitherto unattainable by such means, it also gives a cause for those glacial effects, the history of which is written on the rocks themselves.

Those persons who will examine the simple details of the second rotation will probably perceive that it is merely an accurate description of a movement of the earth, which movement has hitherto been described in a slovenly and vague manner.

During the past two hundred years or more, theorists have been contented with the statement that the earth's axis had a slow conical movement, which caused the pole of the heavens to change its position in the heavens about 20″ annually. Thus this problem was dealt with as though the earth consisted of an axis only. No details have ever been given as to how other parts of the earth moved in connection with this conical movement of the axis.

The pole of the heavens is merely the zenith of the pole of the earth, and that zenith is displaced annually 20″ by this so-termed conical movement of the axis. But are we to be satisfied with this imperfect description, and fail to investigate how the zeniths of other localities on earth are affected by the same movement, which causes the zenith of the pole to move 20″ annually? How much, and in what direction, are the zeniths of 51° north latitude and 70° north latitude displaced annually by "a conical movement of the axis" as regards meridians of six, twelve, and eighteen hours right ascension? It certainly appears remarkable that such important details should never have been examined or referred to, especially when astronomers make such use of the zenith for determining the declination of stars by their meridian zenith distance.

Physical astronomy and the laws of gravitation give no explanation as to why the axis of daily rotation of the various planets differ so much in the angle these make with the planes of the planet's orbits. The axis of Jupiter is inclined about 88° to the plane of its orbit; that of Venus, about 15°; that of the earth, about $66\frac{1}{2}°$; that of Uranus, only about 12°. Neither does physical astronomy explain why the daily rotation of Jupiter and Saturn is more than twice as rapid as that of the earth.

All the sciences that are supposed to bear on astronomy have never yet defined or given any explanation as to how any zenith, except that of the pole of daily rotation, is affected by "*a* conical movement of the earth's axis." This movement of the earth has hitherto been undefined, so that when it is stated that the second rotation of the earth is "not agreed to" by certain supposed authorities of the day, the amusing inconsistency actually exists that these persons do not state what the movement of the earth is with which they do agree.

Whilst the axis of daily rotation is fixed in the earth, the axis of second rotation appears at present not fixed *in* the earth, but is fixed as regards the heavens, the north pole of this second axis being 29° 25' 47" from the pole of daily rotation, and having a right ascension of eighteen hours.

The earth rotates once during about 31,600 years round this axis, in opposition (nearly) to the daily rotation. Thus the conical movement of the two half axes of the earth is *produced* by the second rotation, just as the daily rotation causes a line from the earth's centre to a point on the earth's surface to trace a cone every twenty-four hours.

Exactly as a geometrician can calculate the length of arc over which each locality on earth is carried during a given time by the daily rotation, so can he calculate over what length of arc, and in what direction, each zenith is carried during a given time by the second rotation. We thus have exactness taking the place of that which hitherto has been vague and undefined, with the result that calculations can be made by this knowledge formerly considered impossible.

That this movement should not at once be comprehended

by a geometrician, when proved as it is by calculations, is at least singular, but that certain persons should assert that it is "a vague theory with which they do not agree," whilst they fail to state what it is with which they do agree, is a most remarkable proceeding, and is one which must have been adopted in consequence of the problem not having been by these persons either examined or understood.

That this hitherto undefined movement of the earth, which fully explains the cause of the precession of the equinoctial points, of the change in polar distance and right ascension of the stars, of the decrease in the obliquity of the ecliptic, and also gives the date and duration of the last Glacial Period, and gives the detail changes of every zenith and meridian on earth, should be considered "a mere coincidence," is a statement exhibiting a condition of mind of a peculiar character. Especially is this the case when persons who make this statement assert that an undefined something, called "A conical movement," which fails to explain how any zenith except that of the pole of the earth is affected, which fails to supply them with the means of calculating the polar distance of a star from one observation only for a future date, which fails to give a cause for even the last Glacical Period, is a most profound truth, and is "exact astronomy."

The reader who can free his mind from the dogmatic and contradictory theories which have hitherto reigned triumphant in connection with this movement of the earth, and who will examine the effects of the second rotation, may revel in the number and accuracy of the new problems which he will be able to solve.

The opposition and arguments hitherto urged against the second rotation differ in no way from those brought against the daily rotation of the earth some three hundred years ago. Superstitious theories are put forward, with the firm conviction, apparently, that these can overcome calculation and recorded facts.

In the earlier chapters of this book, proof is given of the power of geometry to demonstrate the form and size of the earth. These proofs are original.

In the later chapters, some of the arguments supposed to be sufficiently strong by their utterers to controvert the second rotation are briefly dealt with.

Many so-called objections which have been advanced are so ridiculously absurd that it would be little short of an insult to an intelligent reader even to call his attention to their puerility.

It may, however, be interesting to collect these objections, and in the future publish these as examples of the class of intelligence which prevailed at the present date, and was supposed to be sufficiently good to set itself up as a guide and instructor of the general reader.

The second rotation of the earth can stand against fair criticism, just as the daily rotation has stood against it. Those persons who consider that personal abuse and the parrot-like repetition of hitherto accepted theories will disprove facts, will occupy in the future a position corresponding to that in which the opponents of the daily rotation now luxuriate.

A. W. DRAYSON.

SOUTHSEA,
   June, 1890.

CONTENTS.

CHAPTER		PAGE
I.	The Form of the Earth proved by Geometry	1
II.	The Movements of the Earth	17
III.	The Known and the Unknown as regards the Earth's Movements	28
IV.	The Second Rotation of the Earth	43
V.	The Precession of the Equinoxes, and the Decrease in the Obliquity	64
VI.	Some Results of the Second Rotation of the Earth	82
VII.	The Pole of the Heavens and the Pole of the Ecliptic	95
VIII.	The Union of Astronomy and Geology	102
IX.	The so-called Proper Motion of the Fixed Stars	113
X.	The Pole-Star	133
XI.	The Nautilus Curve	143
XII.	The Zenith and the Meridian	163
XIII.	The Measurement of Time, and Right Ascension	193
XIV.	Modern Astronomical Observations	220
XV.	The Plane and the Poles of the Ecliptic	229
XVI.	Some Effects of the Second Rotation	241
XVII.	Analogy in the Solar System	255
XVIII.	Objections of Theorists	279

UNTRODDEN GROUND

IN

ASTRONOMY AND GEOLOGY.

CHAPTER I.

THE FORM OF THE EARTH PROVED BY GEOMETRY.

THE student of the history of the progress of astronomical science may probably perceive that, accuracy and truth, have in all past ages been compelled to force their way through the obstruction offered by ignorant superstition, vested interests, erroneous dogmatic theories, and an absence of a correct knowledge of geometry by those persons who were the authorities at various dates.

The success of such obstruction is indicated by the fact that, some five hundred years before Christ, the daily rotation of the earth was taught by Pythagoras and one or two other advanced reasoners, but was rejected and ridiculed by the authorities and their satellites during at least two thousand years, whilst the false theory of the earth being fixed in space was taught in all the schools of astronomy, and the reasoner who denied that the earth was immovable was declared to be not only ignorant of science, but to be actually a heretic.

The oft-repeated error, which has retarded the progress

of true science, has been an indolent habit of hastily inventing theories which might explain some few phenomena, then to cease investigating, or to ignore facts which might prove these theories to be false.

When a theory had been accepted as correct, and taught by authorities during many years, it would cause a considerable amount of trouble if certain facts were accepted which would prove that the theories were erroneous.

The disinclination to re-examine whether any accepted theory is correct has been most marked in all past ages, and it has not been unusual for certain authorities at various dates to merely assert that the then accepted theories were highly satisfactory, in order to check inquiries relative to any novelty (however firmly based on facts) from being even examined.

The value of any theory may be best tested by carrying out computations based thereon. If the calculations can be arrived at correctly by aid of the theory, we may be tolerably certain that we are on firm ground. If, however, facts and calculations must be ignored and denied, in order that the theory should still be imagined correct, we may depend that this theory is untrue.

Among some of the earliest and most ignorant races on earth, it was supposed that the earth was a flat surface bounded by the sky. The learned historian Herodotus, writing on this question, states that "he cannot refrain from laughter when he hears men talking of the earth being round, as though made in a machine."

Such a remark exhibits a mind influenced by prejudice and accepted theories, and also utterly ignorant of geometry.

That the earth cannot be a flat surface, but must be spherical, can be proved by geometry, without any reference to theories or to preconceived opinion.

THE FORM OF THE EARTH PROVED BY GEOMETRY. 3

This proof is to be derived from the following facts. Let us assume for the time being that the earth is a flat surface, and examine what would be the results as regards our satellite, the moon.

We will suppose that three observers are on the earth's surface, and on the same meridian of longitude—one observer, A (Fig. 1), at 60° north latitude; another observer, B, on the equator; the third observer, C, at 60° south latitude. The moon we will suppose to be full, to be vertical at the equator, and on the same meridian as the observers A, B, C.

Now, it is a known fact that, to an observer in 60° north latitude, the moon, when on the meridian and having no declination, has an altitude of about 30° above the horizon. At the same instant the moon would be at the zenith of the observer B, who was at the equator, and would have an altitude of 30° to the observer C.

If the earth were a flat surface, the following diagram would represent the relative positions of the three observers and the moon :—

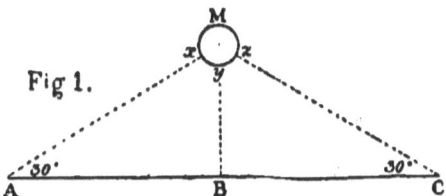

A B C represents the earth's surface, assumed as flat; M, the moon.

The believer in the theory of the earth being a flat surface would claim that here all the conditions relative to the moon's variation in altitude were fulfilled, and consequently that his theory must be correct.

Now let us advance a step further, and examine whether the known facts are explained by this theory of the earth being a flat surface.

Every observer, no matter on what part of the earth's surface he may be, sees the diameter of the moon as nearly as possible of the same size as does any other observer who measures this diameter at the same instant of time.

If the earth were a flat surface the observer at B would be only half as far from the moon as are the observers at A and C. Consequently, the moon would appear to the observer at B, to have a diameter twice as large as that which the observers measured from A and C.

Such change, however, in the moon's diameter does not occur, hence we have a fact occurring, viz. the apparent uniform diameter of the moon, which could not occur if the earth were a flat surface.

Another important fact, however, must be dealt with. When the moon is full, each observer, no matter on what part of the earth's surface he may be situated, sees the same configuration of craters, etc., on the moon's surface. An observer at the Cape of Good Hope will perceive a certain crater on the centre of the moon's surface. An observer at the equator, or in England, will see the same crater in the centre of the moon's surface.

Let us now examine what must be the results if the earth were a flat surface.

An observer at A (last diagram) would see a crater at x in the centre of the moon. An observer at B would see a crater at y in the centre of the moon, and the crater at x on the edge of the moon's limb.

An observer at C would see a crater, z, in the centre of the moon, but he could not see the crater at x. Hence, as an observer travelled about the earth, he would perceive the moon as a very varied object, seeing craters on her surface in the southern hemisphere which could not be seen in the northern hemisphere. Such, however, is not the case; the moon, when full, presents the same "face" to the

THE FORM OF THE EARTH PROVED BY GEOMETRY. 5

observers on all parts of the earth's surface. Consequently it follows that, if the earth were a flat surface, certain facts must occur which are known not to occur; hence the earth cannot be a flat surface.

The strongholds of error and prejudice are not, however, given up without a determined and obstinate struggle, and it has been asserted that if the moon were a disc, in shape like a penny piece, then a crater in the centre of the moon would be seen in this centre, no matter whether an observer were at the equator, and the moon was overhead, or if he were in 60° north or south latitude.

Geometry, however, can again bring to bear a proof which renders this assumption untenable. If the moon seen by an observer, at the equator, and directly overhead, appeared a circle in consequence of being a disc, and not a sphere, then the moon seen at an altitude of 30° by an observer at any other part of the earth would appear as an ellipse, the major axis being twice as long as the minor axis. Such a change, however, in the apparent shape of the moon does not occur. She appears invariably a circle when full, no matter from what part of the earth she is viewed. Hence the assertion that the earth is a flat surface is positively disproved by the appearances of the moon alone.

We now come to the inquiry as to what must be the shape of the earth, in order that the following known conditions relative to the full moon must be fulfilled, viz.:—

1. That the full moon seen by two or more observers at different parts of the earth should be spherical in form.

2. That, no matter whether the moon is overhead or 30° above the horizon, she has always almost the same diameter.

3. That the craters or configurations of the moon always appear the same to all observers, no matter where these observers may be located on earth.

Let A, B, C (Fig. 2), be three observers standing on a spherical body E, and M the moon. To the observer at A, whose horizon would be represented by A x, the altitude of the moon would be represented by the angle x A y, and a crater at o on the moon would appear in the centre of the moon. To an observer at B the moon would be overhead, and would appear very nearly the same diameter that it appeared to the observer at A.

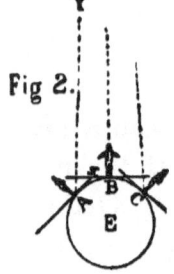
Fig 2.

A crater, o, would appear in the centre of the moon to the observer at B just as it did to the observer at A.

To an observer at C the same conditions would prevail as to the observer at A.

Hence, whilst the theory of a flat earth would cause certain effects, which are known not to occur, a spherical earth explains every known effect. It follows, therefore, that the theory of a flat earth is contradicted by the appearances of the moon, which prove that the earth must be a sphere in order to account for that which every person is capable of perceiving.

Another proof that the earth must be spherical in form is so simple that it seems surprising that it has never been advanced as an unanswerable argument. It is as follows:—

Assume the earth to be a flat surface, such as is shown by the line I E A (Fig. 3), I representing the position of

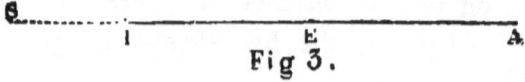
Fig 3.

India, E that of England, A that of America; I being east, A west, of E.

On March 21, the sun rises in the east. Suppose I S to

THE FORM OF THE EARTH PROVED BY GEOMETRY. 7

be the direction of the sun at the time of sunrise. Then it would follow that as soon as the sun was visible from I, it would also be visible from E and A. Consequently it would follow that the sun would rise in India, England, and America at the same instant. We know, however, that this is not the case; we know that when the sun rises in India, it is midnight in England, and when it is midday in India, it is daybreak in England, on March 21. Hence these facts prove that the earth cannot be a flat surface.

Assume that the earth is a sphere, and all the above facts are at once explained.

When we read the remarks of those ancient gentlemen who, like Herodotus, state that they cannot refrain from laughter when they hear men asserting that the earth is anything but a flat surface, we come to the opinion that it is most unfortunate that these gentlemen have not acquired a knowledge of the elementary laws of geometry. Had they done so, they would have been able to prove to their own satisfaction that the earth could not be a flat surface, but must be spherical in form in order to account for known facts.

In the present day geometry is supposed to be taught in our schools and colleges, and a knowledge of this accurate science, it is assumed, is acquired by the students. In too many instances, however, these students learn geometry in a parrot-like manner, and are incapable of applying their knowledge in a practical manner. Examples will be given in the following pages that this statement is not without foundation.

When, however, we find individuals stating that nothing can be proved by geometry, and that it is a useless science, we may be certain that such persons belong to the same class as those who proclaim that the earth is a flat surface, their object being to ignore an exact science which can prove their visionary theories to be not only without foundation, but positively erroneous.

In order to picture in our own minds the geometry of the heavens, we must trace out on the sphere of the heavens the apparent course which the sun, and the various other celestial bodies, appear to trace during twenty-four hours, and during a year. An observer on the earth's surface in, say, latitude 51° N., and who faces the south, should realize that the equator of the earth, if produced to the heavens, forms an arch; the extremities of this arch cut the east and west points of the horizon, whilst the south point of the arch will be 39° above the south point of the horizon. The sun, consequently, on or about March 21 and September 21, rises in the east, attains a midday altitude of 39°, and sets in the west.

If the observer were in north latitude 50° at the above date, the sun would still rise in the east and set in the west, but would attain a midday altitude of 40°.

When the sun is on the equator, a line joining the north and south limb of the sun will always be at right angles to that portion of the "equinoctial" (the term used to describe the trace of the equator on the sphere of the heavens) where the sun is at the time situated.

From this geometrical law we come to one or two interesting items, which at present seem in too many instances to be not only inexplicable, but never to have been observed by certain gentlemen who are regarded as astronomical authorities.

Take, for example, the following diagram (Fig. 4) to represent the horizon and the trace of the equinoctial on the sphere of the heavens from latitude 51° N., E being the east, S the south, and W the west points on the horizon. The sun would rise at E, reach M at midday, and set at W.

We will suppose that when the sun is rising near E

THE FORM OF THE EARTH PROVED BY GEOMETRY. 9

(Fig. 1), a spot is seen on the sun's surface as indicated at Fig. 1. Where will this spot appear on the sun's surface when the sun is near setting, as shown at Fig. 2?

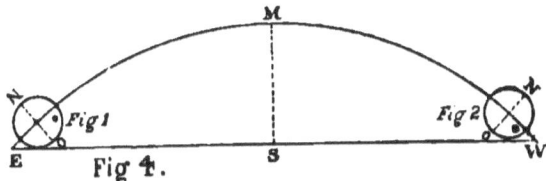

Fig 4.

The reader who wishes to solve this simple problem has merely to draw through the sun's centre a line at right angles to the equinoctial, as shown by the lines N O, to find how the sun's "up" and "down" has changed during the twelve hours it has been above the horizon, and, if the spot has not moved on the sun's surface, it will appear on Fig. 2 in the position shown.

By the rotation of the earth on its axis, the "up" and "down" of an observer is changed; it is not the sun that has altered its position. This apparent change in the position of a spot on the sun, and due to the daily rotation of the earth, can be observed between sunrise and sunset, on any day when there happens to be a spot visible on the sun. The fact, therefore, can be observed, and the cause has been explained in the last few paragraphs.

Another interesting phenomenon, and due to the same geometrical law, takes place with the moon, and is most remarkable when the moon is half full, and is over the equator, or, in astronomical language, has no declination. Under such a condition, the moon would rise in the east, reach the meridian at about 6 p.m., and set in the west at midnight.

To an observer in 51° north latitude, the moon would appear to trace a curve on the sphere of the heavens, the greatest altitude of this curve above the horizon being 39°, when the moon was south, whilst the extremities of the curve cut the horizon at the east and west points.

We will take a date about March 21, when the sun is on the equator, and the moon half full, consequently 90° from the sun.

The line separating the illuminated portion of the moon from the darkened portion will, under the above conditions, be at right angles to that portion of the equinoctial on which the moon is at the time situated. Hence the moon will appear at various hours, during the time she is above the horizon, as shown in the following diagram (Fig. 5):—

E, S, W, represent the east, south, and west points of the horizon, to an observer in 51° north latitude. The curved line represents the trace of the equinoctial on the sphere

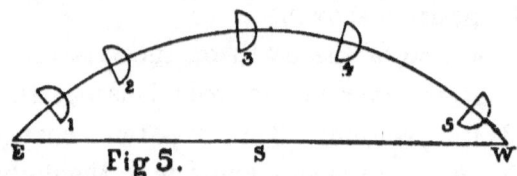

of the heavens. The line separating the illuminated from the dark portion of the moon will always be at right angles to that part of the curved line on which the moon happens to be situated. Thus, shortly after the moon has risen, she will appear tilted as shown at 1. When the moon reaches the south-east, she will appear as shown by 2; when south, as shown by 3; when south-west, as shown by 4.

We may now note a singular fact which leads to some interesting results.

When the moon is at 2, the sun, also on the equinoctial, would be at 4, 90° from 2. And, although we know that the moon shines by the light reflected from the sun, it appears that if we drew a line at right angles to the line

THE FORM OF THE EARTH PROVED BY GEOMETRY. 11

of light and shade in the moon, this line would pass considerably above the sun, not directly on to it. The moon, in fact, appears tilted, and not as though she were illuminated by the sun alone.

If we could see those rays of light which proceeded from the sun and illuminated the moon, those rays would appear curved, as shown in the following diagram (Fig. 6), where S is the sun, M the moon, E the east, W the west points on the horizon :—

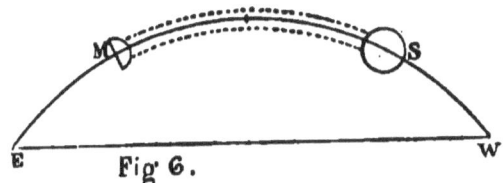

Fig 6.

If an observer were at the equator of the earth, this apparent tilting of the moon would not occur, because the equinoctial traced from the east would pass as a *straight* line to the zenith, and from the zenith would pass as a *straight* line to the west point of the horizon. Hence that portion of the equinoctial between M and S which appears to an observer in 51° latitude as a curved line, would appear at the same time to an observer on the equator as a straight line.

If, again, a comet with a tail 90° more or less extended in the heavens from S to M (last diagram), this tail would appear curved to an observer in middle latitudes, but as a straight line to an observer at the equator.

This simple law of geometry seems to have been overlooked by the learned astronomers of the past, who in several instances have propounded the most verbose and profound theories to explain why a comet's tail should sometimes appear curved, and not straight. Even Sir John Herschel, in his "Outlines of Astronomy," Art. 557, states,

"The tails of comets, too, are often somewhat curved, bending in general towards the region which the comet has left, as if moving somewhat more slowly or as if resisted in their course."

The geometrical law which has been explained relative to the trace of the equinoctial appearing as a great arch in the heavens to an observer in middle latitudes, but as a straight line to an observer at the equator, holds good for a great circle of the sphere cutting the east and west points of the horizon and passing through the pole of the heavens. This great circle would appear as an arch in the heavens to an observer in middle latitudes, but as a straight line to an observer at either pole. This great circle would coincide with the horizon to an observer at the equator, and would to him appear as a straight line.

The following diagram (Fig. 7) represents the trace on the sphere of the heavens of a great circle cutting the east

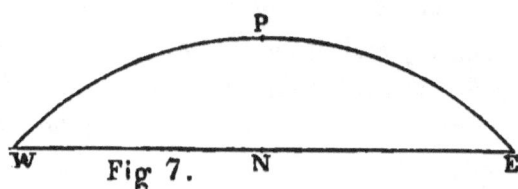

and west points of the horizon, and passing through the pole of the heavens. The observer being in 51° north latitude, and facing the north, W represents the west, N the north, and E the east points of the horizon. P represents the pole of the heavens, the altitude N P being, as a geometrical law, always equal to the latitude, therefore in this case equal to 51°.

The great circle W P E (that part, at least, which is visible) appears to this observer as a great arch in the heavens. If a comet coincided with this great circle, this comet's tail would appear curved to the observer in 51°

north latitude. To an observer at the north pole of the earth, however, the arc E P would appear as a *straight* line, drawn from the horizon to the zenith; the arc P W also a straight line, drawn from the zenith to W. A comet, therefore, which coincided with E P W would appear to the observer at the north pole of the earth to have a straight tail, whereas to the observer in 51° north latitude the same comet seen at the same instant would appear to have a curved tail.

Owing to the rotation of the earth on its axis, causing all celestial bodies to appear to trace circles round the pole of the heavens as a centre, and owing to the fact that an imaginary line drawn from the horizon to the zenith appears as a straight line, it follows that under certain conditions a comet which appeared to have a curved tail would, six hours afterwards, appear to have a straight tail. Such a change during six hours I was fortunate enough to observe with the great comet of 1861.

The geometrical law producing this apparent change is as follows.

Each celestial body appears during twenty-four hours to describe a circle in the heavens round the pole as a centre, therefore during six hours it traces one-fourth of this circle. A line drawn from the horizon to the zenith appears as a straight line to an observer on the earth's surface.

Let us now assume that a comet was seen by an observer in 51° north latitude at 8 p.m., with a tail 90° in length, and coinciding with that great circle on the sphere of the heavens which cuts the east and west points on the horizon and passes through the pole of the heavens. The head of this comet we will suppose 51° from the pole and to the west, the tail stretching to 39° from the pole and to the east. This comet would, under the above conditions, appear to an observer in 51° north latitude as shown below, C P M (Fig. 8). When six hours of the earth's rotation had

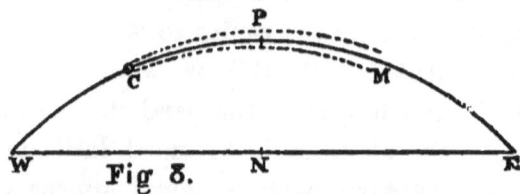

Fig 8.

occurred, that part of the comet at P would not have altered its position in the heavens. The head C would appear to be curved round to N (Fig. 9), the north point of the horizon, whilst the extremity of the tail M would be carried to the zenith of the observer, and the comet's tail would appear to this observer as a straight line, as shown in the following diagram (Fig. 9):—

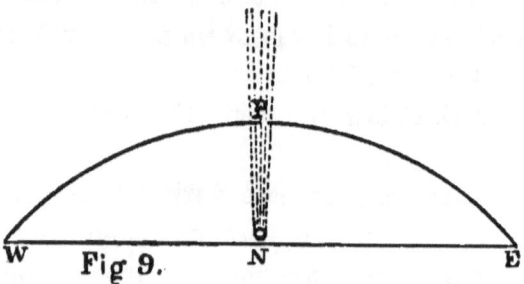

Fig 9.

The comet of 1861 underwent this change during the first night of its appearance. When first seen, the head was near the northern horizon, and the tail stretched in a straight line to the zenith. Six hours afterwards, the head was in the north-east, the extremity of the tail in the north-west, and the tail appeared curved. The cause of this variation is a geometrical one, and is not due to any of those wonderful theories which the imagination of persons unacquainted with geometry have hitherto invented as supposed explanations.

Another example of this geometrical law is afforded by

THE FORM OF THE EARTH PROVED BY GEOMETRY. 15

the "pointers" (a and β Ursæ Majoris) and the pole-star. The pole-star is about 1° 18′ from the pole of the heavens, and during each rotation of the earth appears to describe a circle round the true pole, the radius of which circle is 1° 18′; consequently, at intervals of twelve hours, the pole star changes its position from above to below the pole.

In order to illustrate the effect of the geometrical law relative to apparent curves and straight lines in the heavens the following diagram can be examined (Fig. 10):—

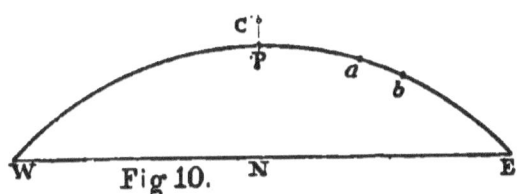

Fig 10.

Suppose W the west, N the north, and E the east points on the horizon, E P W a portion of the great circle of the sphere, passing through the pole P, and cutting the east and west points of the horizon. Suppose a and b two stars on this great circle, and c a star above the pole. The stars a, b, would appear to an observer to point exactly to the star c. When, however, twelve hours of the earth's rotation has occurred, the stars a and b would have been apparently carried round to the position shown in the following diagram (Fig. 11), indicated by b and a; the star, however,

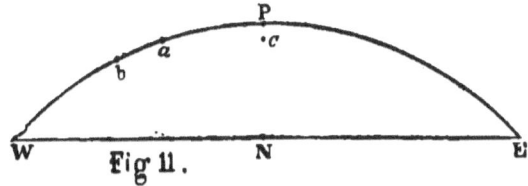

Fig 11.

would by the same rotation be carried to c below the pole, and the stars b, a, would appear to point above the star c, and not directly towards it, as they did on the former occasion.

The "pointers" a and b, Ursæ Majoris, are so situated that when a south-west line can be drawn from the pole of the heavens through these two stars, they will appear to point directly towards the pole-star. When, however, the pointers are to the west of north, they will appear to point *above* the pole-star.

This is a fact as easily observed as is the apparent change in the position of a spot on the sun, or of the tilted moon when half illuminated, or of the curve of a comet's tail, and it is due to the same geometrical law.

It is, however, very remarkable that when, more than thirty years ago, I mentioned these facts to one of the most distinguished astronomers of the time, he stated that he had never remarked them; but, if they really occurred, he should attribute these effects to *refraction*. Had he said that he should have attributed them to electricity or witchcraft, his remarks would have been equally correct.

CHAPTER II.

THE MOVEMENTS OF THE EARTH.

ALTHOUGH the daily rotation of the earth and the annual revolution of the earth round the sun had been accepted as facts by the few advanced minds some five hundred years before Christ, yet the obstruction offered by ignorance and prejudice had prevented these astronomical truths from being generally received until about three hundred years ago, when Copernicus, and afterwards Galileo, revived the theory of the earth's two principal movements.

As soon as the facts could be realized, that the earth rotated every twenty-four hours, and was carried annually round the sun in an orbit either circular or very nearly approaching a circle, the diameter of which exceeded one hundred and eighty million miles, geometricians must have become aware that they were actually residing on an instrument the daily and annual movements of which would enable them to make a survey of the solar system, and of the universe.

That the earth is certainly an instrument, by aid of which we can determine the distance, size, and movements of other celestial bodies, ought to be fully comprehended by the geometrician, in order that he may understand how it happens that there is even now some untrodden ground connected with geometrical astronomy.

The great error that was committed by ancient astronomers was in attributing to other celestial bodies movements which were merely due to the motions of the earth itself. Thus the olden astronomers imagined that the whole of the celestial bodies revolved round the earth every twenty-four hours, whereas it is the earth that rotates every twenty-fours hours. They imagined that the sun moved among the stars, and described a circle among these every year, whereas it is the earth that describes the circle. It thus appears by no means an uncommon error for the mind unacquainted with geometry to attribute movements to external objects which are in reality due to an unknown movement of the earth itself. Having, however, found that in the history of the past this error has been committed, it requires that caution should be used lest we commit the same error, and hastily come to the conclusion that external bodies have some peculiar movement, which after all is due to an unknown motion in the instrument with which we make our observations.

In considering the earth as an instrument with which we make our observations, it is necessary to note certain peculiarities connected therewith. No matter whether the earth be a perfect sphere or a spheroid, we are justified in assuming that the sea-level of corresponding latitudes north and south of the equator is equidistant from the earth's centre. When we examine the distribution of land and water on the earth's surface, we at once perceive that there is an enormous preponderance of land in the northern hemisphere; consequently the centre of gravity of the earth must be north of the equator. There is also a very much larger quantity of land above the water between 15° west longitude and 160° east longitude than there is on the opposite side of the earth. The axis joining the north and south poles of the earth cannot, therefore, pass through the centre of gravity of the earth.

Having, therefore, a rotating and revolving sphere, in which the centre of gravity does not coincide with the centre of the sphere, a mechanician will naturally expect some other movement besides a rotation to take place, however slowly this other movement may occur.

Any apparent movement that takes place in connection with the celestial bodies is readily observed, even by men scarcely superior in intelligence to the savage. The rising and setting of the sun, the change from new to full moon, etc., being examples. To observe these facts is a simple proceeding; to give the true cause for such changes requires intelligence.

Some hundred and forty years before Christ, an observer noted that at the period termed the vernal equinox, and when the sun consequently passes from the south to the north of the equator, the sun coincided with certain stars, with which it did not coincide some hundred or more years previously at the same time of year. In order that the sun should coincide with those stars with which it coincided formerly, it would be necessary to pass the vernal equinox two or three degrees.

Although these ancient observers were unprovided with telescopes, and were unacquainted with such accurate means of measuring small angles as we now possess, viz. verniers and micrometers, yet the size of the instruments used in former times enabled angles to be measured with sufficient accuracy to detect such a change as that just referred to.

When this fact, termed the "precession of the equinox," had been discovered, the ancient observers attempted to explain it as they explained all supposed movements of the stars, viz. by attributing to the sphere of the heavens a slow rotation during a long period of years. The earth at that date being supposed immovable, it would have been considered little short of blasphemy even to hint that the real cause was some movement of the earth, and not of the heavens.

When the daily rotation of the earth and its annual revolution round the sun were admitted as facts, another theory was invented to account for this precession of the equinoctial point.

It was known as a fact that the position of the pole of the heavens varied. The present pole-star, which is now little more than one degree from the pole of the heavens, was two thousand years ago fully ten degrees from this pole; consequently the axis of the earth (the direction of which produced to the heavens determines the position of the pole of the heavens) must have changed its direction. It was perceived that this change in direction of the earth's axis would also explain the precession of the equinoctial point. As at that date it was erroneously assumed that the obliquity of the ecliptic—that is, the angular distance between the pole of the heavens and the pole of the ecliptic—never varied, although the pole of the heavens had altered its position about 10° during about eighteen hundred years, it followed that the pole of the heavens must trace a circle on the sphere of the heavens round the pole of the ecliptic as a centre.

The next step made by theorists was to endeavour to explain what was the movement of the earth which caused the axis to change its direction, and they asserted that the earth's axis traced a cone during a period of about twenty-four thousand years, the pole tracing a circle during the same period on the sphere of the heavens round the pole of the ecliptic as a centre.

Before advancing further in this inquiry, attention must be called to the theory of the axis of a sphere describing a cone.

THE MOVEMENTS OF THE EARTH. 21

When we assert that the axis of a rotating sphere describes a cone, no matter whether this cone is described in twenty-four thousand years or during twenty-four seconds, we must, if we give an accurate description of that which is imagined, state distinctly which pole of the sphere remains fixed, and which pole describes the base of the cone. We must remember that the earth is the instrument with which we make our observations, and every minute movement of this instrument must be known before we can depend upon the observations made thereby. If we imagine that the south pole of the earth remain fixed, whilst the north pole describes the circle or base of the cone, and we happen to be wrong in this guess, we shall be committing the same error that the ancients committed, viz. we shall be attributing to the stars, etc., an independent movement which they do not possess, merely because the instrument by which we determine their positions moves in a manner which has not been suspected.

Hence a great oversight was committed when it was asserted that the earth's axis traced a cone, but it was not mentioned which pole remained fixed, and, remarkable as it may appear to those unacquainted with dogmatic theories, this statement of the earth's axis tracing a cone round the pole of the ecliptic as a centre has remained, during nearly three hundred years, unchallenged by geometricians. Although it has been well known, during the past hundred years at least, that the pole of the heavens was gradually decreasing its distance from the pole of the ecliptic, the supposed centre, yet the various writers on astronomy still continued to repeat that the pole of the heavens traced *a circle* round this supposed centre, from which it was known to decrease its distance; at the same time, it was asserted that the earth's axis traced a cone, although it did not seem to occur to these learned theorists that, if exactitude were desirable, they ought at least to inform the ignorant public which pole it was that remained fixed, and which described the circle.

If a surveyor were supplied with an instrument with which to make observations, and he were told that this instrument rotated on an axis, but that the axis traced a cone, he would certainly like to know whether the upper or lower part of this axis remained fixed, in order that the cone should be traced.

If, also, he were told that the extremity of this axis traced a circle round a point as a centre, from which point it continually decreased its distance, he would be somewhat puzzled to know what sort of circle it could be that thus varied its distance from a supposed centre.

At about the commencement of the Christian era, the obliquity of the ecliptic was found to be about 24°. It is now found to be about 23° 27'. As the obliquity must be of the same value as the angular distance of the pole of the heavens from the pole of the ecliptic, it follows as an interesting inquiry, whether the radius of the circle which it has been stated the earth's axis traces round the pole of the ecliptic has a value of 24°, 23° 27', or some other value. The important geometrical laws connected with the radius of this circle, and the errors which have been made in consequence of the true value of this radius not having been known, will be dealt with in future pages.

The statement that the earth's axis traces a cone, but no mention is made as to which end of the axis remains fixed, is so palpably deficient in detail, that no real investigator or competent reasoner can accept as satisfactory such a vague assertion.

The theory that the earth's axis traces a circle round a point as a centre, from which supposed centre it continually decreases its distance, is so directly in opposition to the well-known laws of geometry, that it is singular how such a belief can be accepted by any person who has learnt even the definitions in his Euclid.

In order that the earth's axis traces a cone, the centre of gravity of the earth must be thrown out of its orbit, as this centre ceases to move round a uniform curve, and unless we know whether it be the north or south pole that remains fixed whilst this cone is said to be described, we cannot state how the centre of gravity of the earth, or any other parts of the earth, are displaced whilst this cone is described.

Now, if, instead of the whole axis of the earth describing a cone, the centre of gravity of the earth remained fixed as regards this movement, and the two semi-axes of the earth described cones, we have every detail of a change in the axis accounted for, although we have very different movements in other parts of the earth.

When, however, we have the two semi-axes describing cones, we have an exactly similar movement occurring during many thousand years that takes place every twenty-four hours, to a line joining the earth's centre with a given locality on the earth's surface. This line traces a cone every twenty-four hours, but the reason why this cone is traced is because the earth rotates every twenty-four hours. If, then, the earth has a second rotation during many thousand years round an axis directed to some point in the heavens, the two semi-axes of the earth would describe cones, and the whole axis would change its direction annually, as it is found to change it, but the centre of gravity of the earth would remain fixed as regards this movement.

In order to comprehend this movement, we will deal with it alone, neglecting for the time being the daily rotation of the earth.

Suppose N S the axis of daily rotation of the earth, C the centre of the earth. Draw an imaginary line, O C P, through the centre of the earth. Keep the points O and P on the earth's surface fixed, and give half a rotation to the sphere representing the earth. When half this rotation has been completed, the half-axis N C will occupy the position N' C; the other half of the axis C S will occupy the position S' C; and, whilst the whole axis will have changed its direction from N S to N' S', every point on the earth's surface will have described half a circle round the two points O and P as centre. There will thus be two points on the earth's surface fixed as regards this movement, whilst the poles and every other point on the earth's surface will describe circles.

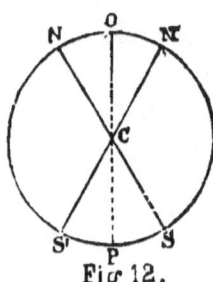
Fig 12.

When a slow second rotation such as the above proceeds very slowly and is mixed up with the daily rotation, and partially concealed thereby, it presents, at first sight, some very complicated movements; when, however, it is dealt with geometrically these apparent complications vanish, and we become acquainted with a mechanical movement, simple and defined, which explains many complications now known to exist. When compared with the present contradictory and vague theory of a circle being described round a movable centre, and an axis tracing a cone when it is not mentioned which end of the axis remains fixed, this second rotation yields most important results.

There appears to be to some minds a great difficulty in comprehending how two rotations can occur at the same time to a sphere such as the earth. Such minds confuse themselves by imagining that these two rotations can be resolved into a third rotation, which contains the other two. Such is not the case. The daily rotation of the earth takes places during twenty-four hours, whereas the second rotation takes upwards of 31,600 years to be completed. The daily rotation is performed round a permanent axis in the earth; the second rotation does not cause this axis to alter its position *in* the earth, but it causes this axis to change its direction as regards external objects.

In order to make the second rotation intelligible to those willing to comprehend it, the following model will be found of service.

Obtain a wooden sphere of any convenient size, through the centre of which drive an iron rod to represent the axis of daily rotation. Let the globe be supported by means of this rod by a circular arc which rests on a spindle and stand, the spindle pointing to the centre of the wooden sphere. The model will then appear as shown in the following diagram :—

The axis of the sphere is represented by N S, the arc by A R, and the spindle by O.

First, give to the sphere a slow movement by causing the arc to turn round. It will be found that

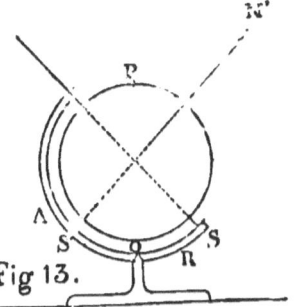

Fig 13.

the point P on the sphere does not alter its position during this movement, but the two semi-axes of the sphere will describe circles, during one complete revolution, round P as a centre. When half this revolution has been completed the axis N S will occupy the position N' S', the two semi-axes having described half a cone. When a complete revolution has been completed, it will be found that every point on the surface of this sphere has described a circle round the points P and O as centres, and, in fact, that the sphere has described a rotation round an axis passing from P to O.

We may now realize the fact that we cannot cause the axis N S to occupy the position N' S' without giving to the sphere half a rotation, and we cannot cause the axis to return to its first position N S without giving to the sphere a complete rotation.

If we now give to the sphere a rapid rotation round N S, and at the same time cause the arc R A to turn round, we have the two rotations taking place at the same time, the one, corresponding to the daily rotation, taking place rapidly; the other, the slow rotation which causes the change in direction of the earth's axis, occurring very slowly.

The second rotation is mixed up and concealed, as it were, by the daily rotation, but it nevertheless exists.

The cause of the precession of the equinox, of the change in direction of the earth's axis, and of other effects, is not, as has hitherto been erroneously stated, a conical movement of the earth's axis in a circle round a point as a centre, from which point it continually decreases its distance, but is really due to a second rotation of the earth.

Immediately the fact is understood that a second rotation of the earth occurs, we have to deal with a problem by no means new. We have to deal with this second rotation in the same manner as we should deal with the daily rotation; we must find in what part of the heavens the pole of this second axis of rotation is situated, a proceeding not fraught with any great difficulty.

Any star which, for the time being, does not vary its distance from the pole of daily rotation is on the arc joining the pole of daily rotation with the pole of second rotation. As the pole of daily rotation is carried round an arc of a circle by the second rotation, various stars will fulfil such conditions, and we have merely to produce the axis joining the pole of daily rotation at various dates with those stars which do not vary their polar distance at various dates, and note where these arcs intersect, and this intersection will give us the position of the pole of the second axis of rotation.

For example, suppose O P Q (Fig. 14) the curve traced by the pole of the heavens during many years. When at O, the pole did not vary its distance from the star x; the pole of second rotation was therefore on the arc O x produced. When the pole was at P, the star y was found not to decrease its distance from P; the pole of second rotation was therefore on the arc P y produced. When at Q the pole was not found to decrease its distance from the star z, and the pole of second rotation was therefore on the arc Q z produced. Where these arcs intersect, as at C, gives the pole of the axis of second rotation.

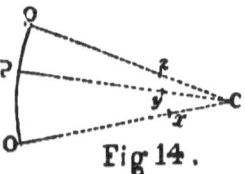

Fig 14.

From an examination of these facts, it will be found that the pole of the axis of second rotation is 29° 25' 47" from the pole of daily rotation, and has a right ascension of 270°.

We can now advance to some most important geometrical laws hitherto overlooked or ignored by theorists. The neglect of these laws has caused endless confusion in the science of astronomy, and has compelled astronomers to employ numbers of observers, night after night, and year after year, in order, from these observations, to frame catalogues of stars.

To a vast multitude of stars, and to the whole solar system, have been attributed independent movements which do not occur, but which appear to occur in consequence of a movement of the earth taking place, which movement has been overlooked.

CHAPTER III.

THE KNOWN AND THE UNKNOWN AS REGARDS THE EARTH'S MOVEMENTS.

ATTENTION will now be drawn to those facts which are at present known to astronomers, and also to those which are not known.

From the observations of the past two thousand years, it is known that the pole of the heavens has a movement of about 20·09" annually, somewhere towards the first point of Aries. It is known that this change in position of the pole of the heavens is due to *some* change in the direction of the earth's axis of daily rotation, and it has been asserted that this change is in consequence of the earth's axis tracing a cone during about twenty-five thousand years. Whether it is the south or north pole that remains fixed, in order that this cone be traced, has not been mentioned.

It is known that this change in the direction of the earth's axis (whatever it may be) is the cause of the shifting of the equinoctial point, termed the "precession of the equinoxes," and of the change in polar distance and right ascension of the stars.

So far we have *approximated* to the cause of known effects, but there is much more to be investigated before we can claim accuracy of detail.

It has been asserted that the earth's axis traces a circle in the heavens round the pole of the ecliptic as a centre.

If this statement were true, then the pole of the heavens must always maintain the same distance from the pole of the ecliptic. The observations of the past two thousand years reveal the fact that the pole of the heavens has, during those years, continually decreased its distance from the pole of the ecliptic. It is, therefore, impossible that the pole of the heavens can trace *a circle* round the pole of the ecliptic *as a centre*, for if it did do so, the two poles would never vary their distance.

It is, therefore, one of the most remarkable exhibitions of mental imbecility to find men, who claim to be astronomers and geometricians, asserting in a parrot-like manner that the pole of the heavens traces *a circle* round the pole of the ecliptic *as a centre*, and yet admitting that these two poles have continued during the past two thousand years to decrease their distance.

If this singular contradiction stood alone it would be sufficiently remarkable; but we have another subject to consider in which mere vagueness assumes to occupy the position of exactitude.

It has been asserted that the earth's axis changes its direction annually to the amount of 20·09″, and this is considered an ample explanation of the mechanical results which occur. Now, if an observer were located at the north pole of the earth, he would find that his zenith at the end of each year occupied a position in the heavens 20·09″ distant from the point it occupied at the commencement of the year.

The north pole of the earth is only one point on the earth's surface, and the zenith of this point changes its position in the heavens annually to the amount of 20·09″, and due to this movement of the earth's axis. How much does the zenith of 51° north latitude, 10° of north latitude, and 70° of north latitude change annually, *and due to the* same *mechanical movement* which causes the zenith of the pole to change 20·09″ annually?

Until geometricians and astronomers can answer these questions accurately, and can define the direction and extent of the movement which takes place annually in the zenith of various localities on earth, they cannot claim to know what really occurs in connection with the known change in direction of the earth's axis.

Theorists have asserted that the joint action of the sun and moon on the protuberant equator of the earth causes the axis of daily rotation to change its direction 20·09″ annually. So completely satisfied do these theorists appear to be with this explanation, that they repeat it as a cuckoo repeats its call, entirely omitting to notice that, whilst they have found from observation that the zenith of the pole of the earth varies its position 20·09″ annually and towards the first point of Aries, no mention has ever been made of how the zeniths of other localities on earth vary annually.

Is the zenith of Greenwich affected in exactly the same manner as is the zenith of St. Petersburg? If not, what is the difference?

We may search in vain for any explanation of how various zeniths are affected by this change in direction of the earth's axis, both in the books written by assumed authorities, or in the Proceedings of the Royal Astronomical Society. We are favoured with numerous theories as to how long the sun will last, and numerous conjectures as to the exact composition of the stars, but it seems beneath the notice of these learned gentlemen to inquire as to how the earth, the actual instrument on which they make their observations, really moves.

It has been an accepted theory during more than two hundred years, that the joint action of the sun and moon produces a change in the direction of the earth's axis, and this theory is considered a sufficient explanation of—what? Of the change in direction of the axis. But how do various zeniths change annually? What is the radius of the circle which the earth's axis traces? Is it 24°, 23°, or something greater or less than either of these values? Such trifling matters seem beneath the notice of philosophers who claim to know how long it will be before the sun is burnt out. Yet upon these items, viz. how each zenith is annually affected, and what is the true radius of the circle which the earth's axis describes, depend the most important results in astronomy.

It has hitherto not been known what the true radius of this circle really is. It has not been known how each zenith is affected, and the result has been that perpetual observation with expensive instruments has been necessary, in order to frame a catalogue of stars for even a few years in advance, and theories invented as to assumed "proper motions" in stars which have no foundation in fact.

Attention will first be directed to the importance of knowing the exact radius of the circle which the earth's axis traces on the sphere of the heavens. Unless this radius be known exactly, it is mere conjecture to talk about the "proper motion" of the stars, and the true radius of this circle has hitherto been a mere vague guess.

There are laws of geometry which are rigid and true, and cannot be ignored; no imaginary theories can set these laws on one side, and the assertions of no assumed scientific authority can prevent these laws from being accurate.

Sir John Herschel, in his "Outlines of Astronomy," Art. 316, states, "It is found, then, that, in virtue of the uniform part of the motion of the pole, it describes a circle in the heavens around the pole of the ecliptic as a centre, keeping constantly at the sâme distance of 23° 28' from it."

In Art. 640 of the same work, Sir John Herschel informs his readers that the pole of the heavens, in describing its assumed *circle*, around the pole of the ecliptic, its assumed *centre*, decreases its distance from this assumed centre 48" per century.

If two such contradictory assertions were made in connection with any simple problem or science, they would be at once ridiculed as too absurd to be meant in earnest, because the asserted conditions are in reality impossible. As, however, the subject dealt with happens to be astronomy, the fact that a geometrical contradiction exists has been overlooked, and it is imagined that a very profound problem is meant when it is asserted that the pole of the heavens describes a *circle* round a point as *a centre*, from which centre the circumference continually decreases its distance.

The theory invented to account for this wonderful movement, which, it must be remembered, is a geometrical impossibility, is that the joint action of the sun and moon on the protuberant equator of the earth causes the earth's axis to *change its direction*. But in what manner does the earth's axis change its direction? In what manner are various zeniths affected, and what is the exact radius of the circle traced by the pole of the heavens? Not one of these important details has hitherto been defined by theorists.

The earth's axis traces a circle in the heavens round the pole of the ecliptic as a centre; at least, so it has been asserted. But as the course which the earth's axis traces decreases its distance from the pole of the ecliptic, the assumed centre, we at once arrive at the important question as to what is the radius of the circle which the earth's axis does trace. Unless the exact radius of this circle can be defined, the assertion that some stars have what is termed "a proper motion" is a mere guess, based on no sound evidence. The reader's special attention, therefore, is called to the following well-known law of geometry (Fig. 15).

Y, P, Q, X, are points on the circumference of a circle, the centre of which is E. The radius of this circle is E P = E Q = E X.

We will now suppose X, Y two stars on the circumference of the circle traced by the earth's axis, and P the position of the pole of the heavens at a given date.

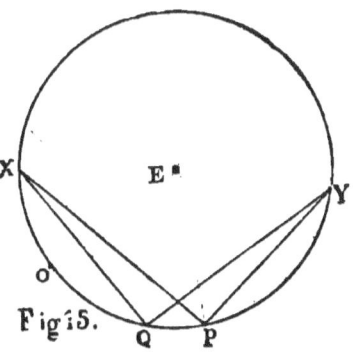
Fig. 15.

The angle X P Y will represent the angle measured at the pole between the stars X and Y, and, in astronomical language, would be the *difference* in right ascension between the stars X and Y.

The well-known law of geometry, to which attention will be called, is that if any points, such as Q, O, etc., be taken on the circumference of the circle Y P X, the angles X O Y, X Q Y, X P Y, etc., will all be equal. In other words, any two stars on the circumference of the circle traced by the earth's axis will never vary in their *difference* of right ascension. The importance of this law must not be overlooked.

Suppose that on January 1, 1800, the difference in right ascension of two stars *supposed* to be on the circumference of the circle traced by the earth's axis was 100°, and on January 1, 1850, the difference was found to be 100° 1'. It would be at once assumed by theorists that these stars had a proper motion, either collectively or individually, of 1' during fifty years.

If, however, the two stars X and Y were not situated exactly on the circle traced by the earth's axis, then they *must* change their difference of right ascension, and this change would not be due to any proper motion in the stars themselves, but to the fact that the radius of the circle which the pole is assumed to trace on the sphere of the heavens had been incorrectly estimated.

Sir John Herschel, in his "Outlines of Astronomy," states that the radius of this circle is 23° 28', and that it never varies; but it is known and admitted that this radius is now about 23° 27' 14", and that about the commencement of the Christian era it was 24°. It is known also, to those persons who are acquainted with the second rotation of the earth, that the radius of this circle is 29° 25' 47", and that the pole of the ecliptic is not, and cannot be, the centre of the circle which the earth's axis traces. It follows, therefore, that to talk about stars having "a proper motion," when the radius of this circle is unknown, and when it is a mere guess as to what this radius really is, affords an example of unreasoning theory, even more remarkable than anything in connection with the past history of astronomy.

We find, however, even in the present day, men passing their time in repeating observations month after month, and publishing tables of the supposed proper motion of the stars (that is, a change in their right ascension or declination), which ought not to occur if the radius of the circle which the earth's axis traces were in reality that which theorists have imagined it to be. That this assumed radius was erroneous, they never seemed to even suspect; consequently confusion prevails, which to such persons is most mysterious, and it requires hundreds of observers to be observing night after night, and numbers of computers to be correcting these observations, in order to frame a nautical almanac for even a few years in advance.

The practical result of not knowing the exact radius of the circle which the earth's axis traces on the sphere of the heavens is, that stars which never alter their relative positions have assigned to them an independent movement, merely because these stars do not vary their right ascensions in accordance with a theory relative to the radius of the circle traced by the pole being of a certain assumed value.

The radius of this circle is not that which has been assumed by theorists, and the endless observations, and the numerous pages of the supposed proper motions of stars, are worthless, inasmuch as the first assumption relative to the course traced by the pole of the heavens is incorrect.

It has been assumed by certain theorists that, when they state that "the joint action of the sun and moon on the earth's protuberant equator" produces a change in the direction of the earth's axis of daily rotation, they have given a complete and profound definition of this movement of the earth, and have also assigned a cause for it.

If any reader imagine that the preceding explanation fully explains this movement, let him take a small globe, and move this globe in accordance with the instructions given above.

In the first place, he must elect which pole of the globe remains fixed, in order that the axis trace a cone, and why should one pole remain fixed any more than the other?

Secondly, he must define how each locality on earth is affected by this movement alone; he must not be contented with the fact that the zenith of the pole changes its position about 20·09″ annually, but he must define how much other zeniths change annually, and also the direction in which these zeniths change, and due to this movement of the earth. Such details are at present untrodden ground to astronomers. We have very elaborate and no doubt profound theories to account for a movement of the earth, but what this movement really is has hitherto never been defined.

In order that the reader may clearly comprehend how vague and indefinite is the present condition of this movement, the following diagram is submitted for examination:—

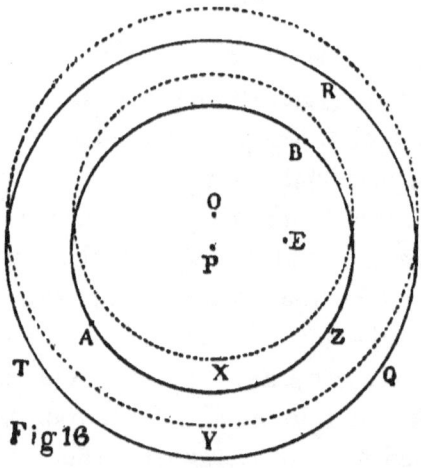

Fig 16

This diagram represents a projection of the northern hemisphere on the plane of the equinoctial. P represents the north pole, Q T R the equator, the circle Z A B a parallel of latitude of 51° north.

We will now take any date in the past when the pole was at P, and when a *daily* rotation of the earth caused the zenith Z to trace a circle round P as a centre during twenty-four hours, viz. the circle Z B A. A point Q on the equator, and on the same meridian of longitude as Z, would also, during one daily rotation, trace a circle, viz. Q R T, round P as a centre.

Any number of years afterwards, the pole is carried to O by that movement of the earth hitherto assumed to be accurately defined in all its details by the term "a conical movement of the earth's axis."

The zenith of 51° is, under these conditions, carried, during a daily rotation, round O as a centre, the circle being represented by the letter X, O X being equal to P Z.

A locality on the equator will, during twenty-four hours, be also carried in a circle round O as a centre, this circle being represented by the letter Y, O Y being equal to P Q.

We now have to consider the effects of this slow movement only, and must treat it as though independent of the earth's daily rotation, and define in what manner the zeniths Z, A, and B, and the zeniths Q, T, and R, have been affected by this movement only.

Whilst the zenith of the pole has been carried over an arc of about 20·09" annually, and in the direction of from P to O, *nearly* in the direction of the first point of Aries, the zenith Z will have been carried over some arc and in some direction; also the zeniths Q, T, A, etc., will have been carried over *some* arcs and in *some* direction by the same movement which has carried the pole from P to O.

The problem to be now solved is, where will the zenith Z be situated at the instant that the pole has reached O, and when a given number of siderial rotations of the earth have been completed.

The zenith Z will have been transferred somewhere on to the dotted circle X; the zenith B will have been transferred somewhere on to the dotted circle X; the zeniths Q, T, and R, somewhere on to the dotted circle Y. But *where* on this circle, is the question.

Unless the exact value, and the exact direction of the arc over which each zenith is carried (and due to the same movement which causes the pole to move from P to O), can be calculated and defined, it follows that the detail movements of the earth are even yet unknown.

All the theories that were ever invented in connection with this change in direction of the earth's axis are valueless, because these theories are supposed to explain *some* movement of the earth, but it has hitherto not been known what this movement really is.

Theories may be venerated as articles of faith, and observations may be repeated by the million, but nothing but confusion can occur unless it can be stated how each zenith on earth, and, consequently, each meridian, changes annually, and due to that movement of the earth which causes the zenith of the pole to move about 20·09" annually.

The amount and direction in which each zenith moves annually has hitherto never been defined by theorists, in spite of the fact that in all observatories the meridian zenith distance of stars is the item measured in order to determine the declination of these stars. That the pole changed its position about 20·09" annually was considered a sufficient explanation, and the changes in each zenith were overlooked.

How each zenith changes, and how important is a knowledge of this fact, will now be explained.

The pole of the heavens changes its direction in consequence of the second rotation of the earth, just as the zenith of a locality on earth changes its direction in consequence of the daily rotation of the earth. The pole of the second axis of rotation is located 29° 25' 47" from the pole of daily rotation, and has assigned to it a right ascension of 18 hours, equal to 270°.

Each zenith describes an arc of a circle round the second rotation pole as a centre, the amount and direction of this arc being dependent on the distance of the zenith from the pole of the second axis of rotation.

With this knowledge, the exact amount and direction in the change of position of any zenith can be simply calculated·

In the following diagram (Fig. 17), P represents the pole of daily rotation at a given date in the past; the circle Z B A, the course traced by the zenith of 51° north latitude during a daily rotation; Q T is the equator; C, the position

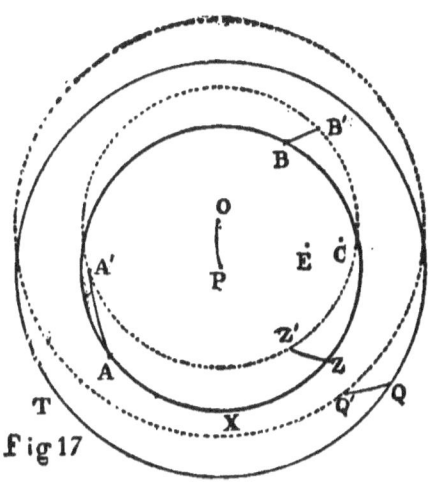

of the pole of the axis of second rotation P C = 29° 25′ 47″; E is the position of the pole of the ecliptic.

The second rotation of the earth is the cause of the change in direction of the earth's axis, and causes the pole, to change its position in the heavens, the pole of daily rotation moving in a circle round the pole C of second rotation. Thus the pole P is carried to O round C as a centre, at the rate (found by observation) of about 20·09″ annually.

The zenith Z will also be carried round C as a centre by the second rotation, and, whilst the pole P is carried to O, the zenith Z will be carried to Z′. In like manner, the zenith Q will be carried to Q′, B to B′, A to A′, etc.

Hence the direction in which each zenith is carried can be accurately defined, and we can advance from the mere vague statement that "the earth's axis has a conical movement," to such details as the direction and amount of movement of each zenith during the year, and due to the second rotation.

In order to obtain the value of the arc over which each zenith is carried annually by the second rotation, we have a very simple problem, the direction in which each zenith moves being at right angles to the arc joining that zenith with the pole of second rotation. The rate at which the second rotation occurs must first be found, and can be obtained as follows :—

A point on the earth's surface, viz. the pole of daily rotation 29° 25' 47" from the pole of second rotation, is carried annually over an arc of 20·09". We have then, C P = 29° 25' 47", P O = 20·09", C A = 90°, to find the value of A B, the arc of the equator of *slow* rotation during one year.

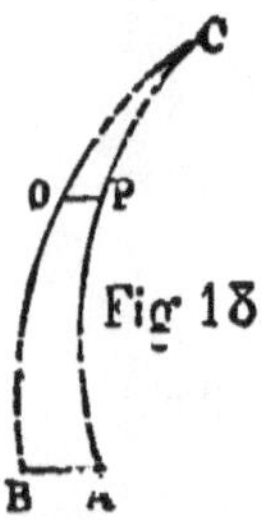

Fig 18.

Making use of the usual formula, O P = A B cosine A P, we obtain 40·9" for the rate of the second rotation annually.

All zeniths which are 90° from the pole of second rotation will be carried annually over arcs of 40·9".

It will be evident that, as each zenith varies its distance (owing to the daily rotation) from the pole of second rotation, the value of the arc traced annually by this zenith will vary considerably, both in amount and direction, according as this zenith is referred to various meridians of right ascension.

It follows also that, as the distance of a zenith from the pole of second rotation will depend on the latitude of a locality on earth, the zenith of which is referred to, the zeniths of two localities will not be similarly affected, either in direction or in amount, if they differ to any great extent in latitude.

Take, for example, two localities on earth, one in 50

north latitude, the other in 60° north latitude, and calculate the value of the arc over which these zeniths are carried by the second rotation when referred to meridians of right ascension of eighteen and six hours.

The zenith of 60° north latitude is 30° from the pole of daily rotation, but is only 30° − 29° 25' 47" = 34' 13" from the pole of second rotation, when referred to a meridian of eighteen hours right ascension. This zenith will, therefore, be displaced annually only to the amount of $\frac{36}{100}$ of a second, found in the following manner: 40·9" multiplied by the cosine of 89° 25' 47" = 40·9" × ·009 = 0·36".

The zenith of 50° north latitude is 40° from the pole of daily rotation, but is 40° − 29° 25' 47" = 10° 34' 13" from the pole of second rotation when referred to a meridian of eighteen hours right ascension. This zenith will be displaced annually by the second rotation to the amount of 40·9" multiplied by the cosine of 79° 25' 47" = 40·9" × 0·183 = 7·3".

Referring to a meridian of six hours right ascension, the zenith of a locality in 50° north latitude will be 40° from the pole of daily rotation, and 69° 25' 47" from the pole of second rotation. This zenith, therefore, will be carried annually by the second rotation over an arc of 40·9" multiplied by the cosine of 20° 34' 13" = 38·3".

A zenith of 60° north latitude, under the above conditions, will be 30° from the pole of daily rotation, and 59° 25' 47" from the pole of second rotation. This zenith will be carried by the second rotation annually over an arc of 40·9" multiplied by the cosine of 30° 34' 13" = 35·2".

Hence the zeniths of two localities, one in 50°, the other in 60° north latitude, will be carried over arcs differing 7" in value when these zeniths are referred to a meridian of eighteen hours right ascension. Yet these zeniths will be carried over arcs annually differing only

3" when referred to a meridian of six hours right ascension.

Can any geometrician or astronomer seriously assert that such important facts as these can be safely overlooked, and that, because a theory is believed in which is supposed to account for a change in the direction of the earth's axis, therefore no further investigations of the true movements of the earth need be undertaken?

The changes in the direction, and the amount of this change for various zeniths due to the second rotation, although hitherto untrodden ground, are more important than all the vague theories ever invented.

CHAPTER IV.

THE SECOND ROTATION OF THE EARTH.

A KNOWLEDGE of how various parts of the earth move in connection with the change in direction of the earth's axis, and consequently the simple means available by aid of which the direction and amount of change annually produced in each zenith by the second rotation, are in themselves important facts, inasmuch as we can substitute accurate details for mere uncertainty. If, however, this were all that could be achieved by a knowledge of the second rotation, it might not be considered of much practical value.

It would be indeed singular if nothing important could be arrived at from a knowledge of this movement of the earth, and it is therefore desirable that some few details should now be given of the calculations that can be made in consequence of, and based upon, the second rotation of the earth.

In the first place, we know that the course traced on the sphere of the heavens by the pole is a circle with a radius of 29° 25' 47", and that the variable circle, with a variable radius asserted to be 23° 28', is an erroneous theory. We know how each zenith is effected by this second rotation, and can therefore calculate every item connected with these changes of zenith. We know, also, that the assertion that a multitude of stars have a large independent movement of their own is erroneous, inasmuch as this statement has been based on the assumption that the pole of the heavens traces a circle round the pole of the ecliptic as a centre, and never varies its distance from this centre, whereas facts prove that, during the past two thousand years at least, the pole of the heavens has not moved round the pole of the ecliptic as a centre, but has gradually decreased its distance from this assumed centre. In other words, an incorrect radius has been imagined for the circle which the pole does trace, and consequently the true course of the pole of the heavens has hitherto been unknown.

If the true course of the pole were really known, no observations would be requisite in order to find what would be the polar distance of stars for any date in the future. This polar distance could be calculated with far greater accuracy than it could be observed, because the errors which may occur, owing to the uncertainty of refraction and of instrumental errors, are avoided.

As an example, the following problem is given for solution by astronomers.

On January 1, 1887, the mean declination of the pole-star was found, by observation, 88° 42′ 21·73″, and its mean right ascension, 1h. 17m. 19·63s. Without any reference to the annual variation in polar distance of this star, but by a knowledge of the true course of the pole of the heavens, calculate the mean north polar distance of this star for January 1, 1850, January 1, 1819, and January 1, 1950.

Those persons who cannot calculate this problem have much to learn in connection with practical astronomy.

One of the first and most important results obtainable from a knowledge of the second rotation of the earth, is the ease with which the position of stars can be calculated for the future, quite independent of the present laborious system of perpetual observation. One accurate observation of the position of a star is sufficient to enable a geometrician to calculate the polar distance of this star for each year for a hundred years in the future or past. For the information of those persons unacquainted with practical astronomy, it may be stated that any such calculation has hitherto been unknown; the method hitherto adopted being to find, by perpetual observation, how much a star increases or decreases its polar distance per year, and then to add or subtract this rate in order to approximate to this star's polar distance for two or three years in advance.

The very fact of such a mere rule-of-thumb system having been adopted proves that the true course of the pole cannot be known; for if it were, a more accurate and less laborious process would be employed.

The means by which such important and hitherto unknown results can be obtained are very simple. From one accurate observation determining the mean polar distance of a star, and its mean right ascension at any given date, we can calculate the distance of this star from the pole of the second axis of rotation, and the angle subtended at the pole of second rotation by two arcs—one drawn from the pole of second rotation to the star, the other drawn from the pole of second rotation to the pole of daily rotation, which is a constant quantity of 29° 25' 47". The third item to be known is the angle at the pole of second rotation formed by these two arcs, a value that can be calculated from one observation of this star. This angle varies, except under rare conditions, at the rate of the second rotation, viz. 40·9" per year—the rate at which the second rotation carries the pole of daily rotation round its circular course. To find the distance, therefore, of the pole of daily rotation from any star becomes a simple problem in spherical trigonometry. The endless observations, therefore, of astronomers, in order to arrive at this result, are quite unnecessary, and those individuals who now pass night after night, and year after year, in measuring the meridian zenith-distance of stars may test whether their observations are correct, and their instruments in adjustment, by means of such calculations as those given below. The observations themselves are quite unnecessary, except perhaps for amusement; but it seems a somewhat expensive amusement to spend many thousands per annum for incomes for those individuals who are employed night after night to make observations, which lead imperfectly to results which can be calculated with the greatest accuracy.

The method for calculating the polar distance of a star from one observation will now be described, and several important stars will be referred to as examples.

The first star to which reference will be made is the pole-star a Ursæ Minoris.

The calculations for this star are (Fig. 19) as follows, C being the pole of second rotation; P, the pole of daily rotation on January 1, 1887; a, the star.

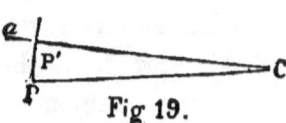
Fig 19.

P C is an arc of 29° 25′ 47″, and the pole P moves round C as a centre at the annual rate of 40·9″.

C a = 29° 52′ 49·6″.

The angle P C a = 2° 27′ 5″ for January 1, 1887.

Now, as the pole P moves round C as a centre at the rate of 40·9″ annually, the distance that the pole was or will be from the star a can be readily calculated in the following manner. Take, for example, the date January 1, 1819. It is required to calculate the mean north polar distance of the pole-star for that date.

Between 1887 and 1819 there are sixty-eight years, during which the second rotation has caused the angle at C

THE SECOND ROTATION OF THE EARTH. 47

to vary at the rate of 40·9″ annually. 40·9″ × 68 = 2781·2″ = 46′ 21·2″. As the date 1819 was earlier than 1887, the angle at C on January 1, 1819, was greater by 46′ 21·2″ than it was on January 1, 1887. On January 1, 1819, the angle a C P was therefore 2° 27′ 5″ + 46′ 21·2″ = 3° 13′ 26·2″. The two sides C a and C P are constants, consequently we have two sides and the included angle of a spherical triangle to find P a the third side, which will be the polar distance of the pole-star for January 1, 1819.

In order to refresh the memory of those geometricians who may not have lately worked out a spherical triangle with the accuracy requisite for such problems as are here dealt with, the detail working of the method of finding the polar distance of this star for a date distant sixty-eight years will be given.

Log. cosine, 3° 13′ 26·2″ = 9·9993121
Log. tangent, 29° 25′ 47″ = 9·7513982

9·7507103 = tan. 29° 23′ 27·3″
+ 29° 52′ 49·6″
− 29° 23′ 27·3″

0° 29′ 22·3′

Log. cosine, 29° 25′ 47″ = 9·9399977
Log. cosine, 0° 29′ 22·3″ = 9·9999842

19·9399819

− Log. cosine, 29° 23′ 27·3″ = 9·9401635

9·9998184 = Log. cos. 1° 39′ 25″ = P a

By this calculation the mean north polar distance of the pole-star for January 1, 1819, was 1° 39′ 25″.

In the Nautical Almanack for 1822, the mean north polar distance of various stars for January 1, 1819, was given; among these the pole-star for that date (1819) was recorded as 1° 39′ 25″.

Thus it is possible, by a knowledge of the second rotation of the earth and of the true course traced by the pole of the heavens, to calculate the polar distance of a star to within a fraction of a second for any number of years in the past or future.

The reader must not jump at the conclusion that the above is a selected case, or "a mere coincidence." Now that the method is explained, he can work similar problems for himself, and test the accuracy of the results.

Below are given results arrived at by calculation, and compared with recorded observation; these facts speak for themselves.

MEAN NORTH POLAR DISTANCE OF THE POLE-STAR.

Date.	Recorded observation.	Calculated.
1887	1° 17′ 38″	1° 17′ 38″
1873	1° 22′ 4·3″	1° 22′ 4·5″
1850	1° 29′ 24·7″	1° 29′ 24″
1819	1° 39′ 25″	1° 39′ 25″
1755	2° 0′ 18·9″	2° 0′ 20″

There is probably no star in the heavens which gives so severe a test of the true course of the pole of daily rotation as does the pole-star. Any person acquainted with geometry must know that the annual rate at which the pole decreases its distance from the pole-star is very variable, whereas with some stars the annual rate is nearly uniform. Consequently, when it is proved that the polar distance of this star can be calculated for a hundred and thirty-two years to within 1″, we may claim that a very severe test has been employed.

It has often been claimed, as a proof of the great labour performed at observatories, that over one hundred observations have been made of the pole-star during the year. There is no denying the greatness of the labour, but the value of this may be questioned when it can be proved that the same results endeavoured to be arrived at by this perpetual observation can be calculated with ease and accuracy.

THE SECOND ROTATION OF THE EARTH. 49

The reader who takes the trouble to work out the details of the second rotation, not only as regards the zenith of various localities, but also as regards the horizon, where the horizon is cut by various meridians, will find some singular changes in various meridians, not one of which changes has hitherto been known to routine astronomers. These changes, in the majority of instances, may cause a very slight apparent change in the rate of the second rotation for some stars, but the changes are due to a geometrical law.

As one of the principal labours of an observatory consists in that perpetual observation with the transit instrument at present considered necessary to obtain the annual rate of increase or decrease in the polar distance and right ascension of stars, so as to frame a catalogue of stars for two or three years in advance, it may be of interest to observers to be supplied with a list of a few stars, and the data by which their positions as regards their mean north polar distance can be calculated with minute accuracy, without any reference to the annual rate of change now found by continued observation.

Such details as regards the pole-star have already been given, and it will be evident that, as the polar distance of this star can be calculated to within $1''$ for a hundred and thirty-two years, repeated observations of this star in order to predict its distance from the pole at future dates, is a proceeding quite unnecessary. So also is it with a multitude of other stars, and as the effect of refraction is always an item of some uncertainty as regards observations, calculation must give more accurate results, when it becomes known how these calculations ought to be made.

The star β Draconis is distant from the pole of second rotation $9° 17' 38''$; the angle at the pole of second rotation, between this star and the pole of daily rotation, was on

E

January 1, 1887, 148° 8' 0". This angle varies at the same rate as the second rotation, viz. 40·9" annually. We have, therefore, a spherical triangle as follows (Fig. 20), C P. From pole of second rotation C, to the pole of daily rotation P = 29° 25' 47". C β, from pole of second rotation to star, 9° 17' 38". The angle P C β, January 1, 1887 = 148° 8' 0"; variation in this angle annually, 40·9".

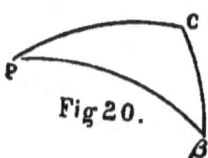

Fig 20.

From the above data we can calculate the distance P β for any date, in the same manner as the mean polar distance of the pole-star has already been calculated, viz. finding the third side when two sides and the included angle are given.

Putting these items in a concise form as follows, they may the more easily be comprehended :—

THE STAR β DRACONIS.

P C = 29° 25' 47" }
C β = 9° 17' 38" } Constants.

Angle P C β, January 1, 1887 = 148° 8' 0"
Annual variation in angle P C β = 40·9"

Calculate the polar distance P β for any other date.

The following results, obtained by calculation from the above data, are compared with the recorded observations at various dates :—

Date.	Recorded observation.	Calculation.
1887 ..	.. 37° 36' 53" ..	.. 37° 36' 53"
1850 ..	.. 37° 35' 8·2" ..	.. 37° 35' 8"
1780 ..	.. 37° 31' 47" ..	.. 37° 31' 47"

The reader who will take the trouble to investigate these facts will perceive that each star will have two "constants" which do not vary, viz. the distance of this star from the pole of second rotation, and the distance of the pole of second rotation from the pole of daily rotation, this last-named distance being 29° 25' 47". The angle

THE SECOND ROTATION OF THE EARTH. 51

formed at the pole of the second axis of rotation by the two arcs above named varies in consequence of the second rotation, but this variation can be calculated, and the angle obtained for any year. Hence the polar distance of the star can be calculated for any year when the above items are known.

The "constants" of a few stars will be found below, by aid of which the polar distance of these stars can be calculated without any reference to their "rate" found by observation, a fact which proves (1) that the time, labour, and expense now devoted to the perpetual observation of these stars is unnecessary; and (2) that, unless the true course of the pole be known, such accurate calculations would be impossible.

When these facts become known to the reader, he will be able to estimate the relative value of the present orthodox theory, and of the second rotation of the earth. He may consider whether the theory that the earth's axis traces a cone round the pole of the ecliptic as the centre, from which centre it continually decreases its distance, is a clear description of how various parts of the earth move in accordance therewith. He may consider whether the theory that "the joint action of the sun and moon on the earth's protuberant equator" is a sufficient explanation to enable him to define exactly how each zenith is displaced during the year by this movement. If he be a reasoner, he may probably inquire why, if the exact movement of the pole of daily rotation and of other parts of the earth be known, it is necessary to spend many thousands of pounds per annum for the income of observers, whose principal occupation is to pass night after night with the transit instrument, in order to find the changes which annually occur in the polar distance and right ascension of stars, so that a catalogue of these stars can be made out for two or three years in advance. These and probably many other similar questions may occur to those persons who look to practical results rather than to dogmatic theories.

The following "constants," and the results obtained from these, are now submitted for investigation:—

THE STAR β URSÆ MINORIS (Fig. 21).

P C = 29° 25′ 47″ }
C β = 21° 50′ 12″ } Constants.

The angle P C β, January 1, 1887 = 31° 34′ 27·3″, rate 40·9″.

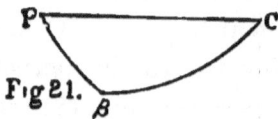

Fig 21.

RESULTS OF POLAR DISTANCE.

Date.	Recorded observation.	By calculation.
1887	15° 22′ 58″	15° 22′ 58″
1873	15° 19′ 32·3″	15° 19′ 31·3″
1850	15° 13′ 53·8″	15° 13′ 53·8″
1819	15° 6′ 17·1″	15° 6′ 17·1″

THE STAR α SPICA VIRGINIS (Fig. 22).

P C = 29° 25′ 47″ }
C α = 89° 46′ 33·3″ } Constants.

The angle P C α = 112° 21′ 9″ January 1, 1887.
Annual rate of variation in angle P C α = 40·9″.

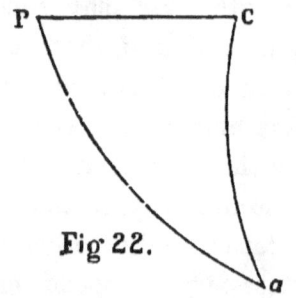

Fig 22.

RESULTS OF POLAR DISTANCE.

Date.	Recorded observation.	By calculation.
1887	100° 34′ 16·6″	100° 34′ 16·6 ′
1850	100° 22′ 36″	100° 22′ 36·5″
1780	100° 0′ 27″	100° 0′ 23″ *

* See Errors of Refraction further on.

THE SECOND ROTATION OF THE EARTH.

The Star α Libræ (Fig. 23).

P C = 29° 25' 47" ⎫
C α = 85° 32' 36·4" ⎬ Constants.

The angle P C α January 1, 1887 = 133° 19' 29·4"
Annual variation in angle P C α = 40·9"

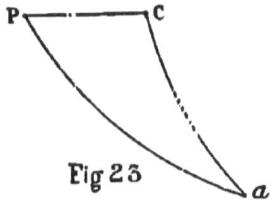

Results of Polar Distance.

Date.	Recorded observation.	By calculation.
1887	105° 34' 17·3"	105° 34' 17·3"
1873	105° 30' 45·3"	105° 30' 45·2"
1850	105° 24' 54·7"	105° 24' 54·7"

The Star Antares (α Scorpii) (Fig. 24).

P C = 29° 25' 47" ⎫
C α = 89° 4' 17·1" ⎬ Constants.

The angle P C α January 1, 1887 = 158° 15' 14·7". Rate 40·9" annually.

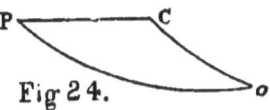

Results of Polar Distances.

Date.	Recorded observation.	By calculation.
1887	116° 10' 49·1"	116° 10' 49"
1850	116° 5' 38·9"	116° 5' 39·3"
1780	115° 55' 36"	115° 55' 39" *

The Star β Libræ (Fig. 25).

C P = 29° 25' 47" ⎫
C β = 77° 5' 20·8" ⎬ Constants.

Angle P C β January 1, 1887 = 137° 1' 46".
Annual variation in angle 40·9".

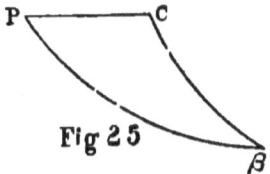

54 UNTRODDEN GROUND IN ASTRONOMY AND GEOLOGY.

RESULTS OF POLAR DISTANCE.

Date.	Recorded observation.	By calculation.
January 1, 1887	81° 2' 5·2"	81° 2' 5·2"
1850	81° 10' 26·9"	81° 10' 27·3"
1780	81° 26' 31"	81° 26' 27" *

THE STAR α' HERCULIS (Fig. 26).

P C = 29° 25' 47"
C α' = 46° 57' 32·7" } Constants.

Angle P C α' January 1, 1887 = 163° 10' 9·3".
Annual variation in angle P C α' = 40·9".

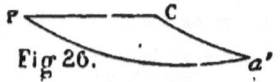

RESULTS OF POLAR DISTANCE.

Date.	Recorded observation.	By calculation.
1887	75° 28' 48·66"	75° 28' 48·66"
1850	75° 26' 4·9"	75° 26' 4·1"
1780	75° 20' 43"	75° 20' 43"

THE STAR α DRACONIS (Fig. 27).

P C = 29° 25' 47"
C α = 26° 37' 4" } Constants.

The angle P C α January 1, 1887 = 54° 45' 21·2".
Annual variation in angle P C α = 40·9".

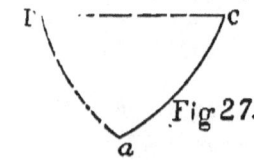

RESULTS OF POLAR DISTANCE.

Date.	Recorded observation.	By calculation.
1887	25° 5' 2"	25° 5' 2"
1850	24° 54' 21"	24° 54' 20"
1755	24° 26' 47·4"	24° 26' 46·3"

* See note on Errors of Refraction.

THE STAR ε VIRGINIS (Fig. 28).

P C = 29° 25' 47" }
C ε = 73° 0' 33·7" } Constants.

The angle P C ε January 1, 1887 = 96° 35' 55".
Annual variation in angle P C ε = 40·9".

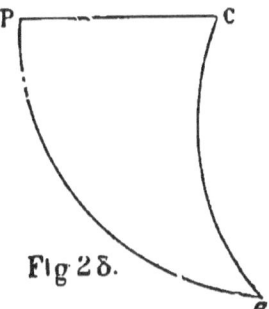

Fig 28.

RESULTS OF POLAR DISTANCE.

Date.	Recorded observation.	By calculation.
1887	78° 25' 59·8"	78° 25' 59·8"
1850	78° 13' 58·3"	78° 13' 58·3"

We have above a few stars whose "constants" are given, by aid of which the polar distance of these stars can be calculated from *one* observation only. No reference need be made to the annual rate of variation in polar distance hitherto arrived at only by long and repeated observation. A true geometrical calculation, based on a knowledge of the second rotation of the earth, enables any person to arrive at results which hitherto have been unattainable. The mere fudge rule of adding or subtracting a certain amount per year in order to give the polar distance of a star for a few years in advance (the only system now practised), is a proof that those who adopt this method are unacquainted with the true movements of the earth, and of the pole of the heavens.

If the true movement of the pole of the heavens were known, the polar distance of a star could be calculated with as great ease and accuracy as we can now calculate the zenith distance of the sun when we know our latitude, the sun's declination, and the apparent solar time.

Observers need not trouble themselves to measure night after night, and year after year, the meridian zenith distances of those stars whose constants have been given, in order that they may find out the rate at various dates at which these stars annually vary their polar distances. Such items can be calculated with much greater accuracy than they can be arrived at by perpetual observation.

When comparing the results as regards the mean polar distance of stars found by calculation with the recorded mean polar distances found by observation some fifty or a hundred years in the past, a cause of error in the ancient records is brought into notice. This error is due to the erroneous value given to refraction by the ancient observers. The following table shows the correction used by the olden observers for various altitudes, and these " corrections " can be compared with the tables of refractions now generally used. As, however, the amount of correction due to refraction varies with the height of the barometer and the thermometer, and also with the amount of moisture in the air, for which latter item no exact allowance can be made, there must and will ever be a small amount of uncertainty even in modern times as regards the true allowance to be made for refraction. Consequently calculation must by reasoners ever be considered more reliable and correct than mere instrumental observation.

The following table will show how varied were the views of ancient observers as regards the value of refraction for different altitudes :—

THE SECOND ROTATION OF THE EARTH.

ALLOWANCE FOR REFRACTION FOR ALTITUDE OF—

	15°.	30°.	45°.	60°.
By La Caille	3' 49"	1' 54"	1' 6"	38"
„ Bradley	3' 30"	1' 38"	0' 57"	33"
„ Halley	3' 17"	1' 32"	0' 54"	31"
„ Sir I. Newton	3' 16"	1' 31"	0' 53"	30"
„ Flamstead	3' 0"	1' 23"	0' 48"	29"
„ Modern tables	3' 35"	1' 41"	0' 58·4"	33·7"

When we find that between even Halley and Bradley there was a difference of 3" for the allowance for refraction for 45° altitude, we need not trouble ourselves much when we find that the calculated polar distance of a star differs from the observed polar distance at the dates 1780 or 1755 even as much as 4"; we leave it to the reader to conclude which is the more likely to be correct, calculations, or observations made with imperfect instruments and corrected by assuming an incorrect value to refraction.

When dealing with the second rotation of the earth, there are several geometrical problems requiring the greatest care before we can correctly calculate results by aid of a knowledge of this movement. Although the polar distance of a star has been referred to, this polar distance is deduced from the meridian *zenith* distance. We have, consequently, to determine how the zenith and how the meridian are affected by the second rotation, before we can make our calculations with certainty. As an example of one among many of these problems, the following case will serve.

It is a geometrical law that the zenith of a locality north of the equator of the earth is carried by the *daily* rotation over a less arc in a given time than is a locality on the equator.

When we refer to the second rotation, this law holds good in its general principles, but the zenith of a locality when this zenith is on a meridian of six hours right ascension, and has a latitude of 29° or thereabouts, will be carried annually by the *second* rotation over *a greater arc* than will the zenith of a locality on the terrestrial equator, the first-named zenith being nearly on the equator *of slow rotation,* whereas the zenith of a locality on the equator of daily rotation being 29° 25′ 47″ from the equator of slow rotation.

Again, when zeniths are near the pole of second rotation, very varied results will occur in the changes of these zeniths produced by the second rotation, both as regards amount and as to their direction. For example (Fig. 29), suppose C the pole of the second axis of rotation, P the pole of daily rotation, A, B, D, E, F, the zeniths of various localities on earth.

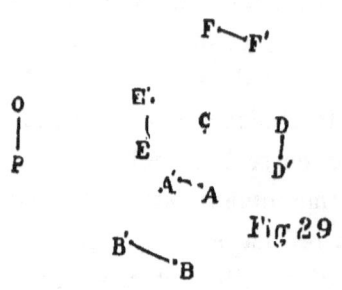

Fig. 29

Whilst the pole of daily rotation is carried from P to O round C as a centre, the zenith B is carried to B′ round C as a centre, A to A′, E to E′, etc., the movements of all these zeniths being round C as a centre. These zeniths consequently differ considerably not only in the direction in which they are carried by the second rotation, but also in the length of the arcs over which they are carried.

In addition to the above-named problems, we have to consider how the equator of the earth is affected at various points by the second rotation, and also how that portion of the meridian between the zenith and horizon is affected by the second rotation.

Those gentlemen who are fully satisfied with the completeness of the theory that it is the joint action of the sun and moon that causes *a conical movement* of the earth's

THE SECOND ROTATION OF THE EARTH. 59

axis, have probably worked out all these calculations. If they have done so, they have kept them secret and have made no practical use of them, for they still continue employing numbers of observers in order to arrive at results for a few years in advance, which results could be calculated for one hundred years in advance with far greater accuracy.

Although the mean rate of the second rotation is $40\cdot 9''$ per annum, the changes produced in the zenith and meridian under certain conditions causes this rate, as regards certain *fixed* stars which have no independent motion of their own, to appear to vary a fraction of a second *of arc* per year.

In order that the reader may understand the small amount in time represented by a fraction of a second of arc when referred to the daily rotation, and hence to the time of meridian transit of stars, he is referred to a table by which arcs are converted into time; he will there see that $10''$ of arc $= \frac{7}{10}$ of a second in time, consequently $2''$ of arc $=$ about $\frac{1}{10}$ of a second in time. Those who may have had considerable experience with chronometers will probably admit that it is rare to find such an instrument which can be relied on for $\frac{1}{10}$ of a second per year. When, however, we come to calculations and to many years, such a minute difference as $\frac{1}{10}$ or $\frac{2}{10}$ of a second of time can be dealt with.

Several important stars will now be referred to, their constants given, and the rate at which the angle at the pole of second rotation appears to vary in consequence of the movement of the zenith and meridian as regards these stars. With the information thus given, the polar distance of any of these stars can be calculated for fifty years or more, without any further aid from observations, and without any reference to the annual variation in polar distance now obtained by observation. In each case the mean polar distance on January 1 of the year named is the item referred to.

60   UNTRODDEN GROUND IN ASTRONOMY AND GEOLOGY.

The Star γ Pegasi (Fig. 30).

P C = 29° 25′ 47″
C γ = 78° 15′ 30·9″ } Constants.

Angle P C γ January 1, 1887 = 81° 8′ 31·7″.
Annual variation in angle P C γ = 40·8″.

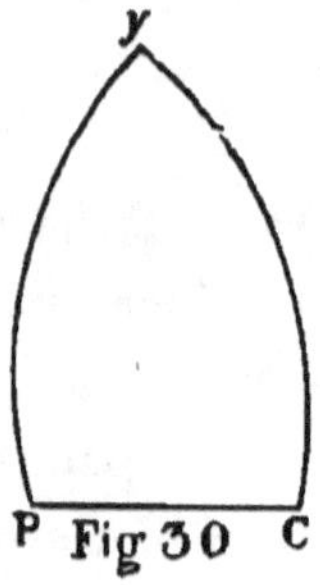

Fig 30

Results of Polar Distance.

Date.	Recorded observation.	By calculation.
1887	75° 26′ 41·1″	75° 26′ 41·2″
1850	75° 39′ 2·1″	75° 39′ 2·5″
1830	75° 45′ 41·8″	75° 45′ 43″
1775	76° 10′ 46·1″	76° 10′ 46·7″

The Star α Aquarii (Fig. 31).

P C = 29° 25′ 47″
C α = 76° 33′ 22·7″ } Constants.

Angle P C α January 1, 1887 = 117° 5′ 32·4″.
Annual variation in angle P C α = 40·8″.

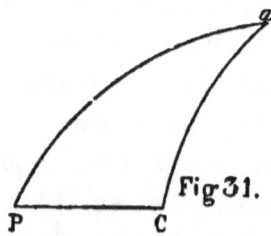

Fig 31.

Results of Polar Distance.

Date.	Recorded observation.	By calculation.
1887	90° 52′ 6·19″	90° 52′ 6″
1850	91° 2′ 47·4″	91° 2′ 47·3″
1780	91° 22′ 55″	91° 22′ 54″

THE STAR γ URSÆ MAJORIS (Fig. 32).
P C = 29° 25′ 47″ }
C γ = 46° 11′ 0·4″ } Constants.
Angle P C γ January 1, 1887 = 53° 49′ 3″.
Annual variation in angle P C γ 40·8″.

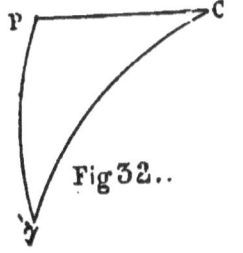

RESULTS OF POLAR DISTANCE.

Date.	Recorded observation.	By calculation.
1887	35° 40′ 37·2″	35° 40′ 37·2″
1873	35° 35′ 57·4″	35° 35′ 57·2″
1850	35° 28′ 16·8″	35° 28′ 16·8″
1830	35° 21′ 35·1″	35° 21′ 36·5″
1755	34° 56′ 35″	34° 56′ 36·7″

These records might be multiplied by the hundred, showing how the polar distance of stars which have no independent motion of their own can be calculated with minute accuracy when the details of the second rotation are known. It may be of interest to watch how many more years official observers will continue, night after night, observing stars in order to obtain the means of framing a catalogue for a few years in advance, when in reality such a catalogue can be calculated independent of any further observations.

In order that the reader may understand how very simple this problem of determining the polar distance of a star becomes by means of a knowledge of the second rotation of the earth, the following items are given, which are sufficient to enable a person capable of working out a spherical triangle to calculate the mean polar distance of a star to one-tenth of a second without any reference to further observation.

THE STAR η URSÆ MAJORIS (FIG. 33).

P C = 29° 25' 47" } Constants.
C η = 36° 31' 5·5" }

The angle P C η, January 1, 1887 = 77° 12' 53"
The annual rate of variation in angle = 40·93"

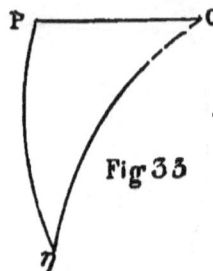

Fig 33.

From the above data the polar distance of this star can be calculated for one hundred years or more, past or future. To continue making observations night after night of this star is much the same proceeding as though a surveyor day after day measured inches on the earth's surface in order to find out how many inches there were in two miles.

Here is another star, viz. α Cephei (Fig. 34):

P C = 29° 25' 47" } Constants.
C α = 22° 58' 32·2" }

Angle P C α, January 1, 1887 = 64° 42' 13·2"
Annual variation in angle P C α = 40·823"

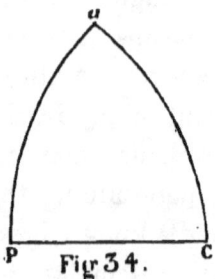

Fig 34.

Find the polar distance of this star for any date, past or future.

The reader, if capable of working out a spherical triangle, is recommended to test these statements for himself; he will, after such tests, not remain in doubt as to whether the polar distance of a star can be calculated to within a fraction of a second, when this star has no independent motion of its own. He will be able to prove that he can calculate this item, and he will realize the fact that those who cannot do so, but are compelled to be perpetually observing, cannot be acquainted with the true movements of the earth.

CHAPTER V.

THE PRECESSION OF THE EQUINOXES, AND THE DECREASE IN THE OBLIQUITY.

WHEN the science of geometry is again taken up, and geometricians realize the fact that dogmatic theories can no more put geometry on one side than these theories can ignore the multiplication table, the fact will become recognized that the statement made, and now accepted as correct and complete, in connection with the precession of the equinoctial point, is one of the most remarkable examples on record, of geometrical contradictions and incomplete reasoning.

We are informed by various writers, who copy one another, that the precession of the equinoctial point produced by "a conical movement of the earth's axis" amounts to 50·1" annually; therefore, write these gentlemen, "as the amount is 50·1" for one year, this is at the rate of 360° for 25,868 years, *which is the period occupied by an entire revolution of the equinoxes.*"

In order to make manifest the errors and unfounded conclusions now prevailing in connection with this problem, attention will be called to the cause which produces the precession, and also the geometrical laws affecting the rate per year.

At the period when the sun's centre is 90° from the

poles of the earth, during the month of March, the vernal equinox takes place. The following diagram will show the course that the earth has followed in order to reach this position, and the cause which produces the precession.

The circle (Fig. 35) represents the earth; N, the north pole; C, the earth's centre; E R, the equator; T C, a portion

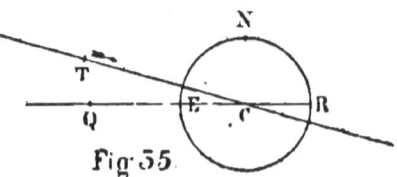

Fig. 35

of the earth's course round the sun, termed the ecliptic. At this period the sun is 90° from the pole N, therefore it is over the equator, the date being when the vernal equinox occurs.

If the axis of daily rotation of the earth were now slightly turned, so that the pole N were moved over towards the sun 20·09″, then the pole N would not be 90° from the sun, but would be 90° less 20·09″.

In order that the pole N should now be 90° from the sun, we should have to move the earth *up* the ecliptic C T, until it reached a point where T Q was 20·09″, and where consequently the sun was 90°, from the pole N.

The distance that we have to move the earth up the ecliptic C T is dependent on three items: (1) the amount of the change in direction annually of the earth's axis; (2) the exact direction in which this change takes place; (3) the value of the angle formed at the time between the plane of the earth's equator and the plane of the ecliptic, technically termed the obliquity of the ecliptic.

When we know the exact direction in which the pole moves, the exact amount of this movement annually, and

F

the exact value of the obliquity of the ecliptic at the date, we can calculate the precession per year for that date; but we must not assume that, because we find this precession of a given value for any one year, we can obtain the whole period by a mere rule-of-three sum. Why we cannot do so will be fully explained further on.

The calculation for obtaining the annual value of the precession, when we know the three items referred to, is very simple, and is as follows. Suppose A B (Fig. 36) the amount of polar movement annually, say 20·09″; A C B the obliquity of the ecliptic, say 23° 28′. The arc A C, which is the arc between two successive vernal equinoxes, can be calculated as follows, A B C being a right-angled spherical triangle :—

$$\begin{array}{ll} \text{Log. sine + Radius A B, viz. } 20'' & = 15\cdot9866049 \\ \text{Log. sine of angle A C B, viz. } 23°\ 28' & = 9\cdot6001181 \\ \hline \text{A C} = 50\cdot2'' = & 6\cdot3864868 \end{array}$$

Consequently, under such conditions we should obtain an annual precession of 50·2″.

Now let us call attention to the important facts in connection with this problem which have hitherto been overlooked.

As long as the pole is carried annually over an arc of about 20·09″, and nearly or exactly towards the first point of Aries, we obtain an annual precession of about 50·2″, as shown by the above calculation. If, however, the radius of the circle which the pole traced in the heavens were only 10°, or if it were 40°, we should obtain exactly the same value, viz. 50·2″ for the annual precession as long as the pole moved 20·09″ annually, and the obliquity was 23° 28′, or very close to this amount. If, however, the radius of the circle which the pole traced were 10° only, this circle would be completed in about 11,270 years found by the formula—

$$\frac{PP'}{\cos. 80°} = EQ$$

If, however, the radius of the circle round which the pole traced its circle of 20·09" annually were 40°, then this circle would be completed in about 43,200 years. Now, the time occupied by the pole of the heavens in completing its circle, is the time occupied by an entire revolution of the equinoxes. This entire revolution is not to be assumed from the rate of the precession at any given date, but is due to, and must be calculated by, the time occupied by the pole in tracing a complete circle in the heavens, and consequently must be calculated on the knowledge of the radius of the circle which the pole really does trace.

Fig. 57.

As this subject is at the present time in absolute confusion, some additional examples will be given, showing the manner in which it can be correctly dealt with, and the geometrical laws that bear upon it.

It is a geometrical law that the arc joining the pole of the earth's daily rotation with the pole of the ecliptic will give a small portion of the arc of what is termed the solstitial colure. This arc being produced each way until it cuts the ecliptic, will indicate those points on the ecliptic where the summer and winter positions of the earth occur.

In the following diagram, T Q I L represents the plane of the ecliptic; E, the pole of the ecliptic; P, the position of the pole of the heavens at a date in the past. P E = 23° 28'.

The arc joining P E and produced to T and I would give the position of the solstitial colure on the sphere of the heavens. The point I on the ecliptic would be the position which the earth would occupy at the period of the winter solstice to the northern hemisphere; T would be the position which the earth would occupy on the plane of the ecliptic at the period of the summer solstice.

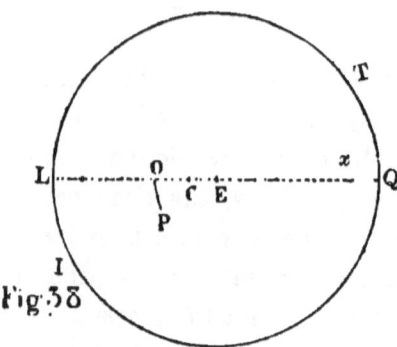

Fig. 38.

Let us now take P O as a portion of the arc traced by the pole of the heavens during many hundred years round the point C as a centre, and let us assume the radius C P of this arc to be only 10° and the pole P to be carried along this arc at the rate of 20″ annually, equal to 1° in one hundred and eighty years.

The movement of the pole of 20″ when the obliquity was about 23° 28′ would, whilst the pole was moving from P to O, and when the obliquity P E and O E varied by only a few minutes, give an annual precession for the equinoctial point of about 50″, found as before described. The course of the pole moving 20″ annually round the small circle of which C, 10° only from P, is the centre, would be completed, as before shown, in 11,270 years.

Because the annual precession happens to be about 50″ at a given date, it is an utterly incorrect assertion to state that therefore the whole revolution of the equinoxes will occupy 25,868 years.

The period during which an entire revolution of the equinoxes occurs depends on the radius of the circle which the pole of daily rotation traces in the heavens, and the annual rate at which the pole moves round this circle.

The annual value of the precession depends on the obliquity of the ecliptic at the date, and on the amount and direction of the polar movement. This annual rate at any particular date will not, however, give us any data by which to calculate the whole period of a revolution of the equinoxes. In order to calculate the whole period, we must know the true radius of the circle which the pole of daily rotation traces on the sphere of the heavens.

In order that this most important fact should be thoroughly understood by the reader, the last diagram will be referred to, and the point x, 40° from P, will be assumed as the centre of the circle traced by the pole P at the rate of 20″ annually. The pole moving from P to O round x as a centre would trace on the sphere of the heavens a very slightly different course during some one thousand years from that which it would trace round C as a centre. This slight difference would be indicated by small variations in the distances P E and E O, which is the value of the obliquity, and also in small variations in the polar distances of some stars.

As long, however, as the obliquity P E varied but slightly from 23° 28′, and the pole moved along its arc at the annual rate of 20″, the annual rate of the precession would be about 50″. But an entire revolution of the equinoxes would, with x as the centre 40° from P, occupy 43,200 years.

When, then, we find it stated by writer after writer who ventures to deal with this subject, that because the annual precession now takes place at the rate of 50·1″ annually, *therefore* it will occupy 25,868 years to complete one revolution of the equinoxes, we may realize the fact that it is very easy to copy errors. Any person, however, who is acquainted with the laws of geometrical astronomy will at once perceive that, as regards this problem, there has been rather too free a use of gratuitous and erroneous assumptions.

The only conditions under which it would be possible to calculate the whole period of a revolution of the equinoxes from the rate found from one year, would be that the pole of the heavens traced a circle round the pole of the ecliptic as a centre, and at a uniform rate, and consequently, as a geometrical law, the distance between the pole of the heavens and the pole of the ecliptic never varied.

The distance between the pole of the heavens and the pole of the ecliptic must be, as a geometrical law, of the same value as the obliquity of the ecliptic. If, then, the pole of the heavens does trace a circle round the pole of the ecliptic as a centre, no variation can occur in the obliquity. We have, consequently, a very simple problem to investigate, viz. whether there has been, during the past two thousand years, any variation in the obliquity of the ecliptic. If there has not been, then the pole of the heavens traces its circle round the pole of the ecliptic as a centre. If there has been any variation, then the pole of the heavens cannot trace its circle round the pole of the ecliptic as a centre, but must trace its circle round some other point as a centre.

It may appear little short of marvellous to the reasoner who is unacquainted with the past history of astronomy, and is consequently unaware of the tenacity with which dogmatic theories, as regards the earth being a flat surface and being immovable, were clung to by the authorities in remote ages, when he realizes the fact that it has been known during more than two hundred years that the obliquity of the ecliptic has been found to decrease, during the past two thousand years at least. This fact has been long known to astronomers, and yet they continue to assert that the pole of the heavens traces a circle round the pole of the ecliptic as a centre. Persons who cling to this belief occupy, relative to geometry and astronomy, the same position as do those who assert that the earth is a flat surface and is not spherical in form, for they consider their preconceived opinions far more sound and true than the rigid laws of geometry.

Recorded observations prove that the pole of the heavens cannot trace a circle round the pole of the ecliptic as a centre, consequently all the theories and calculations based on the belief that it does do so are unsound and incorrect.

The pole of the heavens has decreased its distance from the pole of the ecliptic during the past two thousand years at least, consequently the distance between these two poles has decreased during the same period, and the pole of the ecliptic cannot be the centre of the circle which the pole of the heavens traces in its circular course. Let us now bring to bear on this hitherto confused and contradictory theory, the fact of the second rotation of the earth round an axis inclined to the daily axis of rotation at an angle of 29° 25′ 47″. By this second rotation, it has been already proved that we can calculate the polar distance of a star for one hundred years at least from one observation only, a proceeding hitherto considered impossible by theorists.

This second rotation of the earth will enable those who will take the trouble to examine it, to calculate the value of the obliquity with the same facility by which they could calculate the polar distance of a star, the two problems being, in fact, almost identical.

We find that the pole of the heavens decreases its distance from the pole of the ecliptic. We know the radius of the circle which the pole does trace, and we know the position of the pole of the second axis of rotation; with such data the value of the obliquity can be calculated for any date

The following diagram will show the method of making this calculation, independent of any observations. We can therefore check the observations made at any observatory, and note whether these have been correctly made, instead of being dependent on these, as is now the case.

Let P be the pole of the axis of daily rotation, C the pole of second rotation; P C = 29° 25' 47" (Fig. 39). At the date 2295·2 A.D., the pole P will have been carried to O round C as a centre.

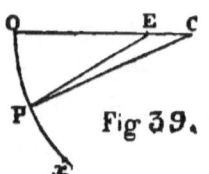
Fig 39.

The angle of second rotation at C varying at the rate of 40·9" annually. E represents the position of the pole of the ecliptic, and at the date 2295·2 A.D., C, E, and O will be on the same meridian of right ascension.

The distance of the pole of the ecliptic E from any given point in the arc x P O will give the angular distance between the pole of the ecliptic and the pole of the heavens, when this latter pole is at that point. Thus E x will be the distance between the pole of the ecliptic and the pole of the heavens when this latter pole is at x, E P their distance when the pole is at P, and so on.

It being a geometrical law that the angular distance between the pole of the ecliptic and the pole of the heavens is always of the same value as the obliquity of the ecliptic, it follows that when the pole was at x, E x indicated the obliquity; when the pole was at P, E P represented the obliquity; and so on.

To find the value of the obliquity for any date now becomes a very simple calculation, inasmuch as C E is 6°, C P = 29° 25' 47", and the angle E C P a variable which can be obtained for any date. We have therefore two sides and the included angle, and we can therefore calculate the third side, which is the obliquity.

For example, suppose we wish to calculate the obliquity for any year, say January 1, 1887.

Subtract 1887 from 2295·2, and we obtain 408·2 years, during which the second rotation progresses at the rate of 40·9″ annually. Multiply 408·2 by 40·9″ and we obtain for the angle at C, January 1, 1887, 4° 38′ 15·38″. With this included angle and the two sides, viz. C E = 6°, and C P = 29° 25′ 47″, the third side P E, the obliquity, can be calculated. This example is worked out in detail below, so that the reader may test other cases for himself.

Log. cosine, 4° 38′ 15·38″ = 9·9985758
Log. tangent, 6° = 9·0216202
―――――――
9·0201960   Log. tan. of first arc = 5° 58′ 49·8″
+ 29° 25′ 47″
− 5° 58′ 49·8″
―――――――
23° 26′ 57·2″ = second arc.

Log. cosine, 6° = 9·9976143
Log. cosine, 23° 26′ 57·2″ = 9·9625649
―――――――
19·9601792
− Log. cosine, first arc = 9·9976298
―――――――
9·9625494 = log. cos. 23° 27′ 14·2″

By this calculation, independent of all observations, we find the mean obliquity for January 1, 1887 = 23° 27′ 14·2″. The mean obliquity for January 1, 1887, recorded in the Nautical Almanac was 23° 27′ 14·22″.

Any other date can be taken and the obliquity calculated quite independently; say, for example, the date January 1, 1800.

Without any reference to the former calculation or to any observations, we can, by a knowledge of the second rotation of the earth, and the true course traced in consequence of the second rotation by the pole of the heavens, calculate the mean obliquity of the ecliptic for January 1, 1800. This mean obliquity is nothing more than the angular distance of the pole of the heavens from the pole of the ecliptic.

74 UNTRODDEN GROUND IN ASTRONOMY AND GEOLOGY.

The whole detail working of finding the obliquity for January 1, 1800, will be given, so that the reader may become conversant with the method, which is merely a repetition of the former example, 1800 being substituted for 1887.

When the reader has examined and knows how to calculate this problem, he will be able to accomplish more in a few minutes than all the observers and theorists have been able to arrive at since astronomy has been treated as a science. It is, therefore, quite worth the expenditure of a little time and thought, to master this simple problem.

To find the mean obliquity of the ecliptic for January 1, 1800.

1800 taken from 2295·2 leaves 495·2 years.

495·2 multiplied by 40·9″ equals 5° 37′ 33·68″, which will be the angle at C, the pole of second rotation for January 1, 1800.

We then have two sides, viz. C E = 6° and C P = 29° 25′ 47″, and the included angle at C, to calculate the third side P E, which is the obliquity for January 1, 1800.

Here are the detail results of the calculation:

Log. cosine, 5° 37′ 33·68″ = 9·9979030
Log. tangent, 6° = 9·0216202
_____
9·0195232 = log. tan. of first arc, 5° 58′ 16·7″

C P = + 29° 25′ 47″
First arc − 5° 58′ 16·7″
_____
23° 27′ 30·3″ = second arc.

Log. cosine, 6° = 9·9976143
Log. cosine, second arc = 9·9625347
_____
19·9601490
− Log cosine, first arc = 9·9976372
_____
9·9625118 = log. cos. 23° 27′ 55·3″ = P E

The mean obliquity, therefore, for January 1, 1800, was 23° 27′ 55·3″. The result arrived at by observers at that date was 23° 27' 55·1". Consequently from 1800 to 1887 the obliquity found by calculation has decreased 41·1". By the results of endless observations it has decreased 40·9", a difference of $\frac{2}{10}$ of a second for eighty-seven years. It may be left to the judgment of the reader as to which is the more reliable, personal and instrumental observation, with the uncertainty always belonging to refraction, or a rigid mathematical calculation.

By calculations similar to those made above, the following results have been arrived at for the mean obliquity :—

Date.	Calculated obliquity.
January 1, 1750	23° 28' 23"
,, 1800	23° 27' 55·3"
,, 1850	23° 27' 30 9"
,, 1900	23° 27' 8 8"

It will be evident to a geometrician that it is impossible that this decrease in the rate can be uniform. There cannot be the same amount of decrease between 1800 and 1850 as there will be between 1850 and 1900. The calculations agree with this law. For example, between 1800 and 1850 there was a decrease of 24·4", which gives a mean rate of 0·488" per year. Between 1750 and 1800 there was a decrease of 27·7", which gives a mean rate of 0·554" per year. Between 1850 and 1887 there was a decrease of 16·7", which gives a mean rate of 0·451" per year.

A knowledge of this variation in the rate of the decrease in the obliquity is most important, especially when the cause of this variation is seen to be a geometrical law. Hitherto theorists have imagined the rate to be constant, a theory impossible according to the laws of geometry, and one that is contradicted by recorded observations.

The reader now possesses the knowledge by which he can calculate the value of the obliquity for thousands of years, without any reference to observations, and he can calculate the rate per year of this decrease with the same ease. For example, suppose we wish to calculate the value of the obliquity for January 1, 890 A.D.; subtract 890 from 2295·2, and we obtain 1405·2 years, which, multiplied by 40·9″, gives 15° 57′ 52·68″ for the angle at C for 890 A.D. Adopting the same calculation as before, and we obtain for the obliquity for 890 A.D. 23° 42′ 48″. By substituting 990 for 890 and calculating as before, we obtain for 890 an obliquity of 23° 40′ 30″, which is at the rate of 2′ 18″ per hundred years for that date.

It is a remarkable exhibition of the small amount of knowledge which has hitherto been possessed on this important problem that theorists should assert that, because the obliquity *between certain dates* was found by repeated observations 48″ for 100 years, therefore they were correct when they assumed it would amount to 480″ for 1000 years, and 4800″ for 10,000 years.

When endeavouring to find what was the value of the obliquity arrived at by observers at distant dates, there are two sources of error which prevent these recorded observations from being quite reliable. The first of these is the erroneous refraction by which the altitude of the sun was corrected; the second is that in ancient times the instruments used were not capable of measuring an angle with greater accuracy than 10′, or at most 5′. In Ptolemy's catalogue of stars, he deals with nothing less than 5′.

As regards the refraction, we will suppose an observer at Greenwich at the date 1690, and he attempted to find the obliquity. In the first place, at that date he did not know the difference between the *mean* obliquity and the apparent obliquity, the effect of the nutation being then unknown. Suppose he measured the greatest meridian altitude of the sun at the summer solstice, and the least meridian altitude at the winter solstice, and, having corrected these altitudes by the table of refractions that were used by Halley or Newton, he made the difference between the greatest and least altitude of the sun 46° 57′ 36″. The half of this amount, viz. 23° 28′ 48″, would be given as the value of the obliquity. The refraction used by Halley for the greatest altitude was nearly 2″ too little, whilst the refraction used for the least altitude was 18″ too little. Hence, if the correct refraction had been used, the difference between the sun's greatest and least altitude during the year 1690 would have been 16″ more, that is, 46° 57′ 52″, and the obliquity consequently would have been given as 23° 28′ 56″, not as 23° 28′ 48″.

It may appear a startling announcement to make to the reader, that he would by calculation discover that an astronomer two hundred years ago either used an erroneous table of refraction, or made an error in his observations amounting to nearly 20″. This announcement is, however, true.

Take the date 1690 from 2295·2, and multiply by 40·9″, and we obtain 6° 52′ 32·68″ for the angle at C. Proceed as in the former calculations, and we obtain 23° 28′ 58″ for the mean obliquity for 1690, within 2″ of what the results ought to be by observation when the correct refraction was used.

The obliquity recorded as having been found by observers about this date, even with their erroneous refractions, are as follows:—

La Hire, 1681	23° 28′ 51″
Picard, 1686	23° 28′ 50″
Wertzelbaur, 1686	23° 28′ 53″

A very excited discussion took place between the French and English astronomers relative to the mean obliquity of the ecliptic for the date 1775. The French astronomers asserted that they had, by the most perfect observations, ascertained it to be 23° 27′ 48″. The English astronomers with equal confidence claimed that they had found it 23° 27′ 59″. M. Le Lande, in his astronomical tables, gives

23° 28' 0" for the obliquity for 1775. The second rotation of the earth and the true movement of the pole of the heavens being then unknown to the mere routine observer, each of the authorities held to their own opinions.

The reader, by the knowledge he has gained from the preceding pages, can prove that the English were more nearly correct than the French, erroneous refraction (probably that used by Bradley) causing them to be only 4" wrong, Bradley using 3' 30" instead of 3' 34" for the correction due to an altitude of 15°.

Taking 1775 from 2295·2, and multiplying by 40·9", we obtain 5° 54' 36" for the angle at C for January 1, 1775, and, working as before, we obtain 23° 28' 3·5" for the mean obliquity for that date; so the error due to refraction of 4" is very creditable to the English observers, for had they used a modern table of refraction they would have come within 1" of the correct obliquity at that date.

An examination of the facts mentioned in the preceding pages will show that the very same movement of the earth which enables a geometrician to calculate the polar distance of a star for each year to one hundred years or more, also enables him to calculate the obliquity of the ecliptic for thousands of years. At the present time no theorist is able to calculate either of these items; he is compelled to depend on repeated observations, and when, by these observations, he has obtained an annual rate for the increase or decrease in the polar distance of a star, he adds or subtracts this rate, and so gives approximately the position of this star for two or three years in advance. By observation, also, he finds approximately what is the annual decrease in the obliquity, and then he subtracts this decrease, and so gives the obliquity for two or three years in advance. No real method of calculation has hitherto been known, because the true movement of the earth has not been known.

THE OBLIQUITY. 79

To add or subtract *a constant* quantity from year to year to the polar distance of a star in order to assign to this star a polar distance for some future date, is erroneous in principle, opposed to the elementary laws of geometry, and would give incorrect results in practice, and it is well known to observers that such is the case. When, however, the decrease in the obliquity is dealt with, a decrease due to exactly the same cause as that which produces the decrease in the polar distance of a star, this decrease is treated as a constant quantity, and from year to year the obliquity is decreased by exactly the same amount, viz. 0·476″, an item extracted from certain solar tables. So utterly unacquainted with the geometrical law relative to the decrease which occurs when a curve approaches a point have the observers of the past proved themselves to be, that they have asserted that, as the obliquity decreased 47″ per century, it would decrease 470″ in 1000 years, and 4700′ = 1° 18′ 20″ in 10,000 years.

Now, the rate in the decrease of the obliquity between 1800 and 1850 was 24·4″, being at the mean rate of 0·488″ per year, but no two consecutive years would have shown the same rate exactly. At the date 1800 the annual rate of decrease was greater than 0·488″. At 1850 it was less than 0·488″. Between 1850 and 1900 the mean rate will be 0·442″.

What are the practical results of adopting this erroneous system of subtracting a constant quantity from year to year, when the quantity is a variable, will be manifest to any geometrician. The error, although small for one year, will go on from year to year until it becomes too large to be ignored; then the accumulated error will have to be wiped out, and a fresh start made. This proceeding was adopted between 1861 and 1865, by altering the assumed rate from 0·457″ per year to 0·557″, and then returning again to 0·476″. Consequently, between the mean obliquity for 1861 and that for 1865 there was made a difference of 2·23″.

The reader can now prove how and why this error occurred. He can calculate the mean obliquity for January 1, 1865, and he will find that between 1850 and 1865 the distance P E decreased 6·9″, which gives a mean rate of 0·46″ per year between these dates. Previous to 1850 the rate was nearly 0·48″ per year. Consequently for many previous years, when only 0·457″ was subtracted, a less amount was taken away than was correct, therefore an adjustment was necessary to set matters right. Between 1887 and 1900 the mean rate will be 0·415″, consequently the present assumed rate obtained from solar tables is slightly too large, and though the present erroneous rate of 0·476″, taken from solar tables, will cause an error of only $\frac{5}{100}$ of a second per year, yet this error, if continued, will in twenty-five years require another adjustment to be necessary. The principle, however, of using a constant quantity for a correction is so palpably erroneous, that it is surprising how such proceedings can be adopted by any person claiming to be acquainted with even the elementary laws of geometry.

It being borne in mind that the rate of the decrease in the obliquity is a variable, just as is the rate in decrease in the polar distance of a star, the reader is not likely to fall into the same error as theorists have hitherto committed. The following rates may therefore be given to show how the obliquity decreases between certain dates:—

Between 1850 and 1865	..	..	..	0·46″ annually.
„ 1865 „ 1887	..	..	..	0·445″ „
„ 1887 „ 1900	..	..	..	0·415″ „
„ 1800 „ 1850	..	..	..	0·488″ „
„ 1700 „ 1850	..	..	..	0·554″ „

These rates cannot be used in order to correct the observed obliquity of any one year for the year following, because these rates will not hold good for any two years in succession. Thus, although the mean rate between 1800 and 1850 may be 0·488", this was not the actual rate at either date. The rate at 1800 was greater than 0·488", and at 1850 it was less. The reader who wishes to find the mean obliquity for any date need not adopt the erroneous system of subtracting a constant quantity from year to year, but he can work out each mean obliquity for any year, not only independent of all observation, but independent of any past recorded obliquity; thus he can calculate what was the obliquity at the date 2000 B.C. as easily as he can calculate what it will be at 2000 A.D., or at any other date.

We may now refer to some of the important matters which can be calculated when we know that the earth has a second rotation, and when we are acquainted with the true movement of the pole of the heavens.

First, we can calculate the varied effects produced by the second rotation on the zeniths of every locality on earth.

Secondly, from one observation only we can calculate the polar distance of a star for one hundred years or more, without any reference to the annual rate found by repeated observation.

Thirdly, we can calculate the value of the mean obliquity without reference to any observations, or any supposed rates now found by observations.

Fourthly, we can calculate the value of the precession of the equinoxes for any year, by first finding the obliquity for that year, and then using the formula given on a previous page.

These results now obtainable with minute accuracy by calculation, have hitherto been arrived at with uncertainty by long-continued and expensive observation.

Several other important matters revealed by the second rotation will be dealt with in future pages.

CHAPTER VI.

SOME RESULTS OF THE SECOND ROTATION OF THE EARTH.

REFERENCE will now be made to some of the methods hitherto considered essential, and the only means known, for obtaining some of the results which can be obtained by very simple calculation. Obtaining the zenith distance of the sun, and also the obliquity of the ecliptic, will first be described.

In a large official observatory the instrument generally used is a large transit instrument, fixed in the plane of the meridian, and being capable of moving in a vertical plane. The latitude of the observatory having been correctly ascertained, the following items become known, viz. the altitude of the pole of daily rotation, which is of the same value as the latitude of the place of observation; the meridian altitude of the equinoctial, which is of the same value as the complement of the latitude.

Hence, if the latitude of the observatory were 51° 28′ N., the altitude of the pole would be 51° 28′, and the meridian altitude of the equinoctial would be 90° − 51° 28′ = 38° 32′.

Each day that the sun crosses the meridian, its zenith distance or altitude is measured by the transit instrument, and this measured distance is corrected for semidiameter refraction, parallax, and any known instrumental errors, and the true zenith distance or altitude of the sun is then obtained, and its altitude above or below the equinoctial will give the sun's declination north or south for the instant at which the observation was made.

In order that the value of the obliquity should be found for any year, it is necessary that we know the greatest distance that the sun reaches north and south of the equinoctial, and as this condition may not occur at the instant that the sun crosses the meridian of the observatory, the following method is adopted. The declination of the sun is measured for each day, for two or three days before and after the day on which the greatest north and south declination occurs, and by a simple calculation the greatest north and south declination can be arrived at; or, in other words, the difference between the greatest and least altitude of the sun during the year is observed. This amount divided by two gives the obliquity. If, then, the difference between the greatest and least altitude were 47°, the obliquity would be 23° 30'; if the difference were 46° 56', the obliquity would be 23° 28', and so on.

The obliquity thus found is not the *mean obliquity*. It is the mean obliquity affected by a small movement of the earth's axis termed the nutation, this nutation completing one complete cycle in about nineteen years. The amount of this nutation is only a few seconds, and, its effects being known, the mean obliquity can be calculated from the observed obliquity for any year.

Practically, therefore, it may be said that the method of finding the value of the mean obliquity is to measure the greatest and least altitude of the sun during the year, to divide the difference by two, and make the necessary allowance for nutation.

The results thus obtained are liable to be affected by the following items:—

First, by personal errors of observation.

Secondly, by instrumental errors.

Thirdly, by the uncertainty of refraction at the time of observation.

Although the probability of the first two causes producing errors may be very slight, yet they exist. The probability of error from the third cause, viz. refraction, is very great even in the present day; whilst fifty and more years in the past, the allowance made for refraction was very incorrect, as has been shown by the different values adopted by different observers. Sir Isaac Newton, for example, in his table of refractions, giving 3' 4" for the summer and 3' 28" for the winter refraction for 15° altitude; whereas, by the tables now in use, 3' 35' is the refraction used for 15° altitude for summer and winter.

It may give an appearance of very great accuracy when accounts are published of official observations, by which it is claimed that altitudes of celestial objects have been correctly ascertained to the one-hundredth of a second. Theoretically this may have been done, but practically it is impossible as long as refraction exists, and there is a liability to the other sources of error named above.

When calculation is employed, only clerical errors, such as incorrect addition or subtraction, or looking out an incorrect log., are possible, and such errors can be at once detected. When these observations are set in opposition to calculation, it is uncertainty struggling against certainty. Fortunately, however, observations have been made with fair accuracy in modern times, and agree very closely with correct results obtained by calculation. Where differences do occur, it can be proved that these arise either from an incorrect use of refraction, or from the theories believed in by observers being opposed to the elementary laws of geometry.

The repeated observations which are now considered necessary, as regards each star, are for the purpose of obtaining an *annual rate* for the increase or decrease in polar distance and right ascension of each star; when this rate is obtained, and the approximate variation in this rate, a star-catalogue for a few years in advance can be framed from the results of these observations. This catalogue, however, is framed merely by adding or subtracting these rates from previous observations; no real calculation based on a knowledge of the true movement of the pole of the heavens has ever been attempted, and for a very good reason, viz. that the true movement of the pole has hitherto been unknown. The reader who has made himself acquainted with the facts dealt with in previous pages, will know that he can calculate the polar distance of a star with accuracy for fifty or one hundred years without any reference to more than one observation of this star.

Having found that, as long as observations fairly reliable can be obtained, the second rotation of the earth and the movement of the pole as herein defined are corroborated, we have merely before us the question of whether we have evidence of nature acting uniformly. It may perhaps be asserted by some theorists that although the earth now rotates from west to east, yet it formerly rotated from east to west. Or that the earth, which now revolves round the sun in a given direction, formerly revolved in the opposite direction. Such assertions are not at all of an unusual character to be made by a certain class of theorist, who delight in wonder-mongering speculations and assertions, in order to attract the attention of the unreasoning. That nature works uniformly is, however, accepted as a fact by the majority of reasoning people, and we may therefore examine what will be the results of one complete second rotation of the earth under conditions similar to those which will now enable us to arrive by calculation at results hitherto considered unattainable except by perpetual observation.

One second rotation of the earth round an axis inclined to the daily axis of rotation at an angle of 29° 25' 47", and proceeding at a rate of 40·9" per year, will occupy, for 360° or one complete rotation, 31,686 years, during which a complete revolution of the equinoctial points will occur. During the half of this period, viz. 15,843 years, the obliquity of the ecliptic will vary 12°, or double the distance of the pole of the ecliptic from the pole of second rotation. The date at which the least obliquity will occur is 2295·2 A.D., when this obliquity will be 29° 25' 47" − 6° = 23° 25' 47"; the greatest was 13,547·8 B.C., when the obliquity was 29° 25' 47" + 6° = 35° 25' 47".

At this latter date, the arctic circle each winter would reach to latitudes of 54° 34' 13" in both hemispheres, and a large portion of England and the whole of Scotland would then have been within the arctic circle.

This condition of the arctic circle would have been similar for both hemispheres of the earth, and there would have been a period of more than 15,000 years during which the arctic circle extended to more than 29° from the poles.

In order that these facts may be clearly understood by those persons who will take the trouble to examine them, a familiar example of the effects of a daily rotation of the earth will be given. Thus during twenty-four hours an observer can perceive exactly similar changes occur in consequence of the daily rotation, that occur during 31,686 years in consequence of the second rotation.

An examination of the following diagram is therefore suggested (Fig. 40).

A projection of the northern hemisphere of the heavens is here given, on the plane of the equinoctial. P is the north pole of the heavens; S, a star distant 6° from the north pole; Z, the zenith of a locality on the earth, which locality is 29° 25' 47" from the north pole of the earth.

Any time of year may be selected for this example, but in this diagram the 21st of March is the date when the day and night are of equal length (equinox).

At midnight the zenith will be situated at Z, and we will assume the star S to be on the meridian and between the zenith and the pole at that hour. The zenith distance of the star S would at that hour be Z S, which is equal to P Z − P S; that is, to 29° 25′ 47″ − 6° = 23° 25′ 47″.

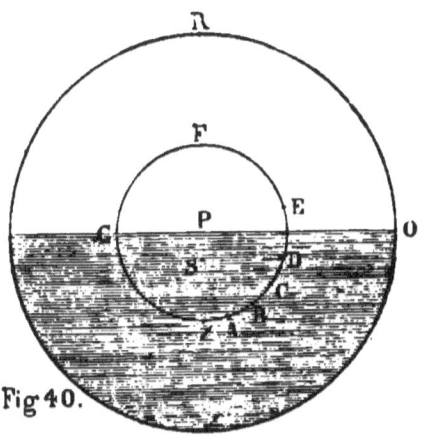
Fig 40.

The effect of the daily rotation is to cause the zenith Z to be carried in a circle round P as a centre during twenty-four hours. Consequently, as 360° of this rotation occupy twenty-four hours, 15° will occupy one hour, and 1° will occupy four minutes of time.

We therefore know the rate at which the angle at P varies, we know that neither the pole P nor the star S alter their position during a few hours, and we know that the zenith Z moves uniformly in a circle round P as a centre, the radius of this circle being always of the same value, viz. 29° 25′ 47″.

Let us now suppose that, two hours after midnight, the zenith has been carried to B by the daily rotation, and that we wish to calculate B S, the zenith distance of the star S.

Two hours of the daily rotation is equal to 30°, consequently the angle S P B is 30°. We know that P S is 6°, and that P B is 29° 25′ 47″. Hence, with the two sides and the included angle, the third side B S can be calculated by the usual formula for spherical triangles. Several examples of the detail working of this problem have been given in the preceding pages.

If at five o'clock a.m. we wished to find the zenith distance of this star, the same process would be adopted, five hours, equal to 75°, being the angle at P for this time.

When the zenith had been carried by the daily rotation to F, no calculation would be requisite. The zenith distance of the star S would then be F P added to P S, equal to 29° 25′ 47″ + 6° = 35° 25′ 47″.

Whilst the zenith was being carried by the daily rotation from Z to A, B, to F, etc., the zenith distance of the star S would increase, *though not at a uniform rate*. Whilst the zenith was carried from F to G, and on to Z, the zenith distance of the star S would decrease, *but not at a uniform rate*.

During any hour of the day or night, however, the zenith distance of the star S could be calculated with the greatest ease, and this calculation could also be made even if the daily rotation occupied 24,000 years instead of twenty-four hours.

During the night, however, when the stars are visible, this change in the zenith distance of a star can be observed, and those observers who care to test this problem may do so by using some instrument, and measuring the zenith distance of a star, and comparing the results obtained by observation with those arrived at by calculation.

It will be evident to the most average intellect that, when a rotation occurs round a fixed point, the figure described must be a circle; consequently each zenith during twenty-four hours of the daily rotation must describe on the sphere of the heavens a circle, the common centre of all these circles being the pole of the heavens.

Self-evident as this fact no doubt appears to the reasoner, or person gifted with common sense, yet attention is called to the fact for reasons which will be evident when the opinions of theorists as regards this problem are referred to.

From the daily rotation of the earth and the movement of a zenith produced thereby, we will now advance to the second rotation of the earth, and the movement of the pole of the heavens caused thereby during one second rotation. The two problems are almost identical, and the results can be calculated with the same ease. The only difference being that the second rotation is partly in opposition to the daily rotation, and that we must treat the pole of the daily rotation as we treated a zenith in the former example. The daily rotation takes place during twenty-four hours; the second rotation takes place during 31,686 years.

In the following diagram (Fig. 41), C represents the pole of the second axis of rotation; E, the pole of the ecliptic, 6° from C; P, the position to which the pole of the heavens will attain at the date 2295·2 A.D.; L M N O P the course over which the pole of the heavens has been carried since the date 13,547·8 B.C.

When the pole was at L, the distance to L E, which would be the measure of the obliquity, was C L = 29° 25' 47", added to C E, which was 6°; equal, therefore, to 35° 25' 47".

The rate at which the pole is carried round this circle by the second rotation is 31,686 years for 360°. One-fourth of this circle, viz. from L to M, would occupy about 7921 years. Consequently, at the date about 5626 B.C. the pole was at M, and the distance E M was about 30°, which gives the value of the obliquity at that date.

In 7921 years more the pole would be carried to P, at which date, viz. 2295·2 A.D., the obliquity of the ecliptic will be 29° 25′ 47″ − 6° = 23° 25′ 47″. Thus from the date 5626 B.C. to 2295·2 A.D. the obliquity will have decreased about 6½°, but it will not have decreased this amount at *a uniform rate.*

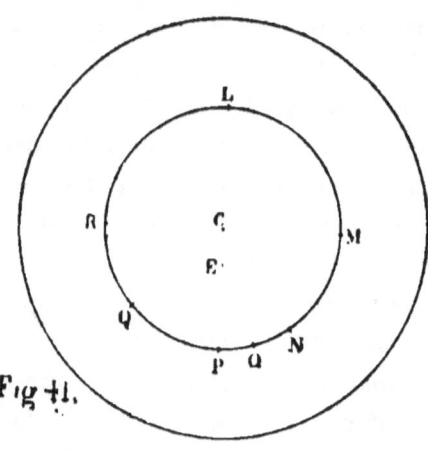

Fig. 41.

Between M and N, the decrease was rapid; between O and P, the decrease was small. To find the value of the obliquity for any date, we proceed exactly in the same manner as when we wished to find the zenith distance of a star for any hour of the night or day. We may select any date, say 500 A.D., for which we wish to know the obliquity. We subtract 500 from 2295·2, and multiply the remainder by 40·9″, and thus obtain the angle at C for the date 500 A.D.; then with the two sides, C E = 6° and C N = 29° 25′ 47″, and the included angle at C, we can calculate N E, the obliquity; N, we will assume, representing the position of the pole of daily rotation at the date 500 A.D.

From an investigation of this problem, any geometrician will perceive into what serious errors observers will fall who imagine that, by subtracting a constant quantity annually from the mean obliquity found by observation at any date, they can assign a true value to the obliquity for a future date.

The difference in rate is small between different years at present, as already shown—thus between 1800 and 1850 the mean rate annually was − 0·488″, and between 1850 and 1865 it was − 0·46″ annually—but it is a serious error to imagine that the mean obliquity for 1000 or 2000 years in the past can be arrived at merely by adding 46″ or 48″ per century to the mean obliquity found by observation for 1889. This problem cannot be dealt with in this mere rule-of-thumb manner, but must be treated like any other geometrical problem.

The fact that the second rotation of the earth, as herein described, causes a variation of 12° in the obliquity of the ecliptic, and consequent extension of the arctic circle, may probably be of more interest to the general reader than other portions of the subject.

An extension of 12° in the arctic and antarctic circles means a vast change of climate on earth, particularly affecting middle latitudes, although but slightly influencing the climate of tropical regions.

From the date about 21,000 B.C. to about 6000 B.C., an arctic climate would prevail down to 60° latitude in each hemisphere during winter; whereas a tropical climate would prevail in summer in each hemisphere up to 30° at least. The midday altitude of the sun at the height of this period, viz. at about 13,000 B.C., would have been 12° greater during each summer, and 12° less during each winter than it is at present.

Every person who has travelled in northern regions, or who has even crossed the Atlantic during summer, knows the effect of the sun's heat on arctic regions. It is during summer that icebergs are liberated and float southwards carrying their burthens of boulders, etc., which are deposited as the icebergs ground and melt. It is during summer that the masses of snow are melted, and vast floods of water rush over the country. Add 12° to the midday altitude of the sun during summer, and subtract 12° from the midday altitude in winter, and the effects which now occur, moderately only, in high northern regions, would have occurred with far greater force down to 54° latitude in each hemisphere.

These conditions were comparatively only recent in the past history of the earth. They lasted from about 21,000 B.C. to about 6000 B.C., viz. whilst the pole was carried by the second rotation from R to L and M, last diagram.

It seems at least singular that, whilst the second rotation of the earth enables any person acquainted with it to calculate with the greatest accuracy results considered by modern theorists unattainable except by means of perpetual observation, this same movement, if continued during one entire second rotation, should prove that such a climate and such conditions must have prevailed on earth just previous to historic times.

It is so singular, because geologists have during years asserted that their facts proved that just previous to historic times some such change of climate must have prevailed on earth in order to account for known effects.

Modern theorists have denied that any change of climate can occur from astronomical causes, except of a very minute character. Their reasons for this denial will be examined further on. The reader may then be able to form an opinion as to the value of such assertions, and to really weigh the evidence advanced by such theorists, and to place this in the scale opposite to that in which the multitude of facts herein given may be placed.

Attention may, however, be directed to the facts which

are brought into notice in this book. The observations of the past two thousand years prove that the earth's axis has continually changed its direction during these years. The rate of this change being very slow. An arc of only about 25° is the arc whose character is presented for analysis. The examination of this arc reveals the fact that it is the arc of a circle having a radius of 29° 25' 47", and that the pole of daily rotation traces this circle in consequence of the earth having a second rotation. From a knowledge of this movement, the polar distance of a star can be calculated for one hundred years or more, a proceeding hitherto unknown in astronomy. A knowledge of the same movement enables us to calculate the precession of the equinoctial point with minute accuracy for any date. It also enables us to calculate the value of the obliquity of the ecliptic for any date, and shows that just previous to historic times there was a period of 15,000 years at least during which the arctic circle extended its limit from 6° to 12° more than at present, thus effecting vast changes in the climate of middle latitudes on earth. The same movement enables us to calculate how each zenith on earth is affected yearly and at various times of the year. Many other important results depending on this movement will be described in future pages.

By means of the present orthodox theories, the polar distance of a star cannot be calculated for any remote date, a fudge rule of adding or subtracting a certain value found by observation being the only method hitherto known. The value of the obliquity of the ecliptic cannot be calculated, the only means at present known being to subtract *a constant* quantity annually from the obliquity found by observation, with the result that it is necessary to fudge the records occasionally, to keep facts and theories in agreement.

It is asserted by theorists that no change greater than 1° 21' can occur from astronomical causes in the extent of the arctic circle, and consequently that geologists must look elsewhere than to astronomical science for an explanation of their facts.

CHAPTER VII.

THE POLE OF THE HEAVENS AND THE POLE OF THE ECLIPTIC.

WHEN individuals capable of reasoning possess the moral courage to free their minds from scientific dogma and the influence of mere authority, and realize the fact that astronomy, like all other sciences, must be dealt with by aid of common sense, of the exact laws of geometry, and by the careful investigation of facts, they may then perceive that the manner in which two problems in astronomy have been dealt with by theorists is most remarkable.

These two problems are the changes or possible changes in the obliquity of the ecliptic, and the so-termed proper motion attributed to nearly every star in the heavens.

The first of these two problems will now be dealt with.

When geology, after being ridiculed and opposed, had at length established itself as a science, it became acknowledged that the facts with which geologists were acquainted, plainly indicated that in former times climates in certain latitudes had existed quite different from those which now prevail, and which have prevailed, during the past 2000 years.

Among the most remarkable and interesting of these changes, because probably it was the most recent, and its evidence the most easily seen, was the great boulder period, or glacial epoch. The evidence existing throughout Europe and North America, and also in the southern hemisphere down to about 50° latitude, of an arctic climate having prevailed just previous to historic times, was convincing. The evidence even showed certain details. It showed that this climate came on gradually; that the arctic climate crept down from the north, reached a maximum, and as slowly and gradually retired, until the present climatic conditions prevailed. There was distinct evidence that icebergs carrying enormous boulders, which now are liberated each summer in present arctic regions only, were formerly liberated in localities as far south as the north of England and corresponding latitudes in Europe and America.

There was evidence of vast quantities of *fresh* water having flooded the country and produced those deep beds of gravel and sand now spoken of as " the drift," as though snow and ice in abundance were periodically melted, and thus produced these periodic floods with their consequent results.

It was not the boulders and drift only which caused geologists to become convinced of the great change of climate which had occurred in comparatively recent times in middle latitudes; but the deposits of the flora and fauna in the drift was a mixture of an arctic and almost tropical character, as though the climate were sometimes arctic, and sometimes almost tropical.

These facts being well known to geologists, they appealed to astronomers to ask whether there might not be something in astronomy which had been overlooked or incorrectly interpreted, and which might aid to give a solution of those mysteries, which had hitherto been admitted as unsolved problems in a grand science.

This demand from geologists to astronomers was by no means unreasonable, and it would have had a firmer base

THE INCLINATION OF THE AXIS. 97

had geologists been better acquainted with some of the known facts connected even with observational astronomy.

The planet Venus, which is the nearest planet to the earth, revolves round the sun in an orbit which is inclined to that of the earth at an angle of about 3° 23' only. Yet the axis of daily rotation of Venus is inclined to the plane of its orbit only 15°. Hence the arctic circle, and an arctic climate, prevails each winter in Venus down to within 15° of her equator, whilst during the summer of each hemisphere a tropical climate prevails to within 15° of her poles.

The orbit of the planet Uranus is inclined to that of the earth at an angle of about 46' only, yet the axis of this planet very nearly coincides with the plane of her orbit. Consequently the arctic circle in this planet extends nearly to the equator, and the tropics extend nearly to the poles. The most varied changes of climate must prevail, therefore, on Uranus during its long year of about 30,686·8 mean solar days.

The planet Jupiter moves round the sun in an orbit which is inclined to that of the earth at an angle of about 1° 19' only, but the axis of rotation of this planet is so nearly vertical to the plane of its orbit round the sun, that scarcely any change of climate takes place on Jupiter during its year. The variations from summer to winter which are experienced on earth cannot occur on Jupiter, the length of the day during midsummer scarcely varying from the length of day during midwinter, a uniform climate prevailing during each year in every latitude on that planet.

When, then, geologists appealed to astronomers for some solution of the strange variations of climate revealed by geology, they ought to have been aware that the course or orbit which a planet during its year followed round the sun, was not a very important item as affecting the

H

annual changes of climate in a planet. The important question should have been, what were the changes in the direction which the axis of this planet went through, or might go through, as regards the angle which it made with its orbit round the sun?

How was it that whilst the orbit of the planet Venus differed only 3° 23′ from that of the earth, yet her arctic and antarctic circles reached to within 15° of her equator? How was it that whilst the orbit of Jupiter differed only about 1° 19′ from the orbit of the earth, yet the arctic and antarctic circles on Jupiter did not extend even 3° from her poles?

The question to be investigated was, what probable or even possible movement might take place in a great rotating sphere, which might cause its axis to make varied angles with its course round the sun? Was astronomical science acquainted with any change in the direction even of the earth's axis? If so, the subject for investigation was plainly indicated.

The facts of geologists were numerous and strong, and naturally they appealed to astronomy for a solution, more especially when, as was probably known at the time, the condition of other planets was very different, as regards their annual changes of climate, to those which have prevailed on earth during the past 2000 or 3000 years.

The problem for investigation was one of great interest, and requiring very careful geometrical handling. It was, in reality, to define the nature of the curve which the earth's axis had traced on the sphere of the heavens during the past 2000 years of which we have records. The manner in which this pole varied its distance from certain stars, especially from those near it, such as the pole-star, λ Ursæ Minoris, ϵ Ursæ Minoris, β Ursæ Minoris, etc., would give co-ordinates by which the nature of this circle might be ascertained, and the problem was one well suited to a geometrician.

That it was possible for two curves quite different in their general nature to be somewhat similar during a small portion of their course, the following diagram will prove.

Suppose the circle L I T (Fig. 42) to be the plane of the ecliptic; E, the pole of the ecliptic; P, the pole of the axis of

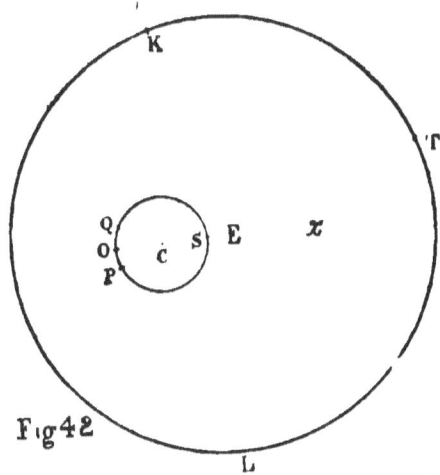

daily rotation, say 2000 years ago; C, the centre of the circle round which the pole P might be carried in the direction P O Q S.

Let E P = 24°, C P = 11°.

The pole, in its movement from P to O at the rate of 20″ annually, would decrease its distance from E; there would consequently be a slight decrease in the obliquity owing to this movement, because a decrease in the obliquity is the same thing as a decrease in the distance of the pole of the heavens P from the pole of the ecliptic E.

There would be a precession of the equinoctial point in consequence of this movement of the pole from P to O. There would be a decrease in the polar distance of the stars in the direction towards which the pole was moving, and an increase of their polar distance in the opposite point of the heavens. The small arc P O, if 10° in length, would occupy, for the pole to travel, some 1800 years at the rate of 20" annually.

The curve from P to O, being a portion of the arc of a circle having C for its centre, would differ but little from an arc of a circle having x for its centre. E x we will take, for example, as 6°; there would be a difference which a geometrician could easily discover, although a slight one.

A very vast difference, however, would occur in the general results, according as C or x were the centre of the circle which the earth's axis described.

If C were the centre, then, when the pole P had been carried round half its circle to S, the poles E and S would be only 2° apart. The obliquity of the ecliptic would then have been 2° only; the arctic and antarctic circles would have extended only 2° from the pole; the tropics would have extended only 2° from the equator; and an almost uniform climate during summer and winter would have prevailed in every locality on earth.

With x as a centre, and a radius of about 30°, P, when half its circle had been described, would have been 36° from E. The arctic and antarctic circles would then have extended 36° from the poles, and the tropics 36° from the equator, and extreme cold in winter and extreme heat in summer would have prevailed in middle latitudes on earth.

As the question raised by geologists was one relative to the changes of climate which facts proved had occurred at remote periods, the inquiry as to the cause was narrowed into a single item. It was to investigate what had been the course of the pole of the heavens and the movement of the earth in recent times; what was the curve traced by the pole of the heavens; what was the radius of this curve; was the curve a circle? if so, where was its centre? Conse-

INCLINATION OF THE AXIS.

quently, to determine the true character of this movement of the earth was the real problem submitted for solution.

It was a known fact that there was nothing impossible in a planet which rotated daily round an axis and revolved annually round the sun, having its axis inclined only 15° to the plane of its orbit, as was proved by the planet Venus. There was nothing impossible in a planet having its axis inclined about 88° to the plane of its orbit, as was proved by the planet Jupiter. In fact, the angle which the daily axis of rotation made with the orbit along which the planet travelled, presented every variety from 90°, as shown by Jupiter, to 2° or 3°, as shown by Uranus.

These variations were not due to the orbits of these planets being inclined to each other at very different angles, for it was known that these orbits differed from that of the earth in two of the cases mentioned less than 2°. Consequently, this great variation in the angle which the planet's axis made with its orbit must be due to some movement or cause in the planet itself, not in its orbit round the sun.

These facts being known, it will be evident in what manner the problem submitted by geologists ought to have been investigated; and it will now be stated how it was, and has been treated.

CHAPTER VIII.

THE UNION OF ASTRONOMY AND GEOLOGY.

When geologists had brought pressure to bear on astronomers relative to the change of climate which might occur from astronomical causes, it was decided that the problem should be submitted to M. La Place, the French mathematician and theorist, for his investigation and decision.

The manner in which M. La Place examined the case was as remarkable as it was inappropriate.

Instead of endeavouring to investigate and define the true curve traced by the pole of the heavens over an arc of 10°, of which we have correct and approximate evidence, according as we refer to modern or ancient observations, M. La Place set to work to discover how much the earth's orbit round the sun might vary.

Instead of testing by the rigid laws of geometry, such as the change in polar distance of stars near the pole, what the true character of this curve really was, and if the curve were a part of a circle, giving the true radius of this circle, M. La Place, by theory, announced that the plane of the ecliptic (that is, the earth's orbit round the sun) could not vary from a mean quantity more than 1° 21'. What this variation (assuming that it really occurred) had to do with the problem submitted by geologists, it is difficult to imagine.

If the plane of the ecliptic varied as much as 3° 23', then the earth's orbit and that of Venus would be in the same plane. Yet the arctic circles in Venus would even then reach more than 70° from her poles, whilst the arctic circles on earth would differ but slightly from their present limits.

Having arrived at the conclusion that the orbit of the earth round the sun could not vary from a mean position (what the mean position was, this profound theorist did not consider worth mentioning) more than 1° 21', he made the astounding assertion, that *therefore* the obliquity of the ecliptic could never vary more than 1° 21', and *therefore* *exact* astronomy could give no help to geologists as regards any variation of climate in the past, as the limits of extension of the arctic circle *must be* 1° 21' from this "mean."

Before mentioning the results which followed this so-termed investigation and its announcements, we must refer to the interesting assumption which must have been made by this theorist before he could put forward such assertions as those he made.

A change of 1° 21' in the plane of the ecliptic means the same thing as a change of 1° 21' in the position which the pole of the ecliptic occupies in the heavens.

That a change of 1° 21' in the position of the pole of the ecliptic should cause a change of 1° 21' in the obliquity, it must follow as a geometrical law that the pole of the ecliptic must either move away from, or come nearer to, the pole of the heavens by 1° 21', because the distance between the pole of the heavens and the pole of the ecliptic is the exact measure of the obliquity.

In order that this decrease of 1° 21' should follow, as a matter of course, a change in the position of the pole of the ecliptic of 1° 21', the pole of the heavens must trace a circle round some centre, from which centre the pole of the ecliptic varied its distance 1° 21'.

Where this centre was, and how far from the then position of the pole of the ecliptic, was considered not worth investigating or mentioning.

Strange to say, however, whilst the assertion above mentioned was put forward as a finite and unanswerable truth, this same gentleman stated that the pole of the heavens always described *a circle* round the pole of the ecliptic *as a centre*.

If the pole of the heavens did describe *a circle* round the pole of the ecliptic *as a centre*, no variation whatever in the position of this centre could cause a variation in the distance of the circumference from its centre. The circumference itself might present some peculiar curve on the sphere of the heavens, but the distance of the circumference of a circle never can vary its distance from its centre, no matter how much this centre alters its position as regards other objects.

We can slowly describe, with a pair of compasses, a circle on a piece of cardboard whilst we are travelling in a railway train. The centre of this circle will alter its distance from surrounding objects, but it will never alter its distance from the circumference of the circle of which it is the centre.

When, then, M. La Place, having investigated what might be the change in the position of the plane of the ecliptic as regards the fixed stars (an inquiry which had nothing to do with the problem submitted to him), arrived at the conclusion that, as the pole of the ecliptic could vary only 1° 21' from some imaginary mean position, therefore the obliquity could vary only 1° 21', he made an assertion so utterly incorrect and devoid of any foundation in fact, that it is almost inexplicable how such an error could have been overlooked even sixty years ago.

If the centre of the circle which the earth's axis traces

be 6° from the pole of the ecliptic, we must have a variation of 12° in the obliquity without any change whatever in the plane, and hence in the pole, of the ecliptic itself.

No sooner, however, had this erroneous statement of the theorist been made known, than hundreds of followers, with a blind submission and an unreasoning adherence to authority, joined in a chorus to the effect that it had been proved that no changes of climate ever had occurred on earth, and that *exact* astronomy was, therefore, unable to give any aid to geology.

When we reflect on the particular problem that was submitted to the French theorist—the facts that were at his disposal, and the evidence presented by geologists, and by the condition of other planets, and the actual movement of the earth which was known to occur—and then find that his investigation was limited to an inquiry relative to the plane of the ecliptic, whilst he utterly ignored the movements of the earth itself, we believe that every reasoner must admit that it is one of the most remarkable exhibitions of dogmatic unreasonableness ever put before the scientific world.

The result, however, was to stop inquiry in this direction. Geologists, convinced of their facts, were hurrying along what appeared to them a clear and open road, by which they could reach the waters of truth, when suddenly M. La Place blocks the way, and "no thoroughfare" in this direction is the announcement. Checked in their attempt to reach truth by aid of astronomy, geologists seemed to lose their common sense, and immediately invented a series of extravagant and baseless theories, evolved from their imagination, and in which they were aided by wonder-loving lecturers and others, in the endeavour to account for their known facts.

After the announcement had been made that theory

proved that the plane of the ecliptic, "by the joint action of all the planets," could not vary more than 1° 21' from some imaginary mean position, and the consequent opinion arrived at, that no matter how the earth moved, yet no change greater than this 1° 21' ever had or ever could occur, it was considered a proof of ignorance for any person even to hint at any greater change having taken place, even during millions of years.

It was not considered in the least unscientific to assume that the earth, which, it was asserted, had once been a red-hot globe of fire, had cooled down till it had become so cold as to produce the glacial epoch, but was now warming up again.

It was stated that the sun, which, it was asserted, was a great blazing body, kept going by shrivelling up, or by fuel in the form of meteors being shovelled into it, must have become hard up for this fuel, and the fire consequently had nearly gone out, and thus caused the cold of the glacial period; but that now, from probably having been stirred, it was again burning brightly.

In order to leave no ground uncovered by speculations, it was also asserted that the whole solar system was rushing through space with enormous velocity, and therefore probably in this rush passed through climates which might be very hot or very cold, just as a traveller who might travel from the equator to arctic regions would encounter great changes of climate.

These, and a multitude of other wonder-mongering theories, were considered really scientific probabilities; but to imagine that any change more than 1° 21' in the obliquity could or had ever occurred, was stated to be a proof of ignorance.

Had not the great French theorist, by the most exhaustive analysis, proved that the *plane of the ecliptic* could vary only 1° 21' from some mean position, and, no matter how the earth moved whilst travelling round this ecliptic, there could not by any means be produced a difference in the obliquity of more than 1° 21'?

The course which the pole of the heavens traced was thoroughly well known to theorists. It was a circle traced during 25,868 years round the pole of the ecliptic as a centre, from which centre it never varied its distance.

How any variation in the obliquity could occur if the earth's axis did trace a circle round the pole of the ecliptic as a centre, was considered too trifling a matter to be even discussed.

Divesting this problem of the verbiage and contradictions by which it has long been surrounded by theorists, a few practical facts will now be brought into notice.

In order that any theorist can positively affirm that during all time there can be no change greater than 1° 21' in the obliquity of the ecliptic, he must know with minute accuracy the following items :—

First, the exact course which the earth's axis has traced and will trace on the sphere of the heavens during millions of years must be known. If this course be a circle, the exact radius of this circle must be known, the exact position of the centre of this circle, and whether this centre ever has, or ever will alter its position.

Second, *the detail movements of the earth must be known, which take place in connection with the change in direction of the earth's axis.*

These two items are of the first importance.

Third, what change has or will take place in the earth's orbit round the sun, by which change the position of the pole of the ecliptic will be altered, and the results produced by the first-named movement might be slightly modified.

When M. La Place asserted that, because the plane of

the ecliptic, by his theories, could vary no more than 1° 21', and that *therefore* the obliquity could not vary more than 1° 21', he must have imagined that he knew with minute accuracy those items mentioned as first and second in the preceding paragraphs. If he did know these items, we have at once brought to our notice a neglect which was little short of unpardonable.

If the true course of the pole on the sphere of the heavens had been (as it was claimed) correctly known, then, as this pole was carried uniformly round its known curve, the distance of this pole from various stars could be calculated for any date, without reference to observations. If the course of the pole had been correctly known, not for millions of years (as was asserted), but only for one hundred years, then no further observations need be made in order to calculate by simple geometry the polar distance of a star for one hundred years, and without any possibility of error.

Here was an opportunity for M. La Place and his followers to give a practical proof of the correctness of their theories. They might have said, "Observations are liable to error, both from personal and instrumental causes; refraction is also a very fruitful source of error. Thousands of pounds are spent annually at various observatories for the incomes of observers, who sit up night after night, with their eyes at a transit instrument, in order to find out the rate, and the variation in the rate, at which the pole moves towards, or away from, a star. All this labour and expense is unnecessary, because, we having by theory proved that the plane of the ecliptic can vary from a mean position only 1° 21', everything as regards the movements of the earth must be known, and consequently to continue observing in order to find out what is known, would be as childish as to employ men to count how many times the second-hand of a chronometer went round, during twenty-four hours."

Had the assertions of these gentlemen been facts instead of mere theories, they might have given such proofs as the following:—

"As we know the true course of the pole of the heavens, we will prove we know it. We will take the following stars near the pole, and demonstrate that by our knowledge we can, from one observation only (and without any reference to the variable change in polar distance hitherto found by aid of repeated observation), show how to calculate the polar distance of these stars for each year, for one hundred years or more. Here are a few stars only: a Ursæ Minoris, λ Ursæ Minoris, ϵ Ursæ Minoris, β Ursæ Minoris, a Draconis, β Draconis, λ Draconis, and many others."

M. La Place and his followers, however, gave no such practical proof that their theories were correct. They were fully aware of the immense amount of time and labour given to observations, to say nothing of the expense, all of which must be totally unnecessary if, as they asserted, they knew with exactitude the true course of the pole of the heavens. Yet they did not give the slightest hint of their profound knowledge, which, if possessed, would have rendered further observation quite unnecessary.

Here in the present day, scores of years after this wonderful discovery, men are employed night after night to observe stars, in order that they may ascertain that which, it was claimed, was known with minute accuracy, and could therefore be arrived at by calculation, without any further reference to observation.

Surely this was an unpardonable piece of neglect on the part of the French theorist and his followers. To possess such perfect knowledge, and yet to continue carrying on laborious observations as though no such knowledge were really possessed, was an exhibition of remarkable inconsistency.

Let the reader refer to Chapter IV. of this book, and he will see that the polar distance of the stars given, all near the pole, can be calculated with minute accuracy for each year for one hundred years or more, without any reference to observations.

No matter whether the annual variation in polar distance of these stars be great or small, be variable or almost constant, yet by the same geometrical laws, can their mean polar distance be calculated either for past or present years.

This calculation can be effected only when a knowledge of the true movement of the pole, and of the second rotation of the earth, is possessed by a geometrician.

Hitherto theorists have not known either of these facts, and hence even in the present day there is necessity for perpetual observation.

When, then, M. La Place asserted that, because the plane of the ecliptic could vary only 1° 21', therefore the obliquity could vary only 1° 21', he made a statement utterly devoid of any foundation. When he stated that astronomy could give no help to explain the facts of geology, he made another statement devoid of truth. Astronomy can and does explain the facts of geology, but it is not the present orthodox astronomy of those theorists who assert that the centre of a circle can vary its distance from the circumference, and yet always remain the centre of the same circle. Nor is it the astronomy which asserts that no variation in the obliquity can occur, except by a change in the plane of the ecliptic. The planets Venus and Jupiter tell a different tale, even if geometry were unknown.

So completely did the theory of M. La Place stop all inquiry relative to the cause of geological climates being due to movements of the earth, and so completely have theorists since then been as it were mental slaves to the assertions that were then made, that more may have been written on this subject than it really deserves, these errors and oversights being almost self-evident.

Yet the history of astronomy proves how the strongholds of error are obstinately clung to by incompetent reasoners, and when we read the remarks that have been made by certain authorities relative to the second rotation of the earth, it is not difficult to understand how it was that during many hundred years, the daily rotation of the earth was rejected and sneered at, by the astronomical authorities of the past.

Some years ago, when standing on the banks of a lake in Nova Scotia (a locality well suited to the study of the evidence of the glacial period), I observed that the hard rocky shore was cut and marked by the glaciers and icebergs of the boulder period. In various inland localities were enormous boulders, which had been carried many miles from the parent rocks, and deposited in what was now a vast forest. My only companion was Paul, a Micmac Indian.

Pointing to the boulders and the marks on the rocks, I said, "Paul, how do you account for all this?"

Paul, without any hesitation, replied, "Long time ago, more winter in winter, more summer in summer. More winter make more snow, more icebergs; more summer melt snow quicker, float icebergs more than now. That what I think."

This Indian hunter was so ignorant of science that he did not know even the multiplication table, yet his opinion was correct.

When we compare the reasoning of this son of the forest with some of the assertions of modern scientists, it really seems possible, that the mere sing-song mutterings of the village softy may contain more truth and real science, than the long-incubated opinions of the over-crammed, dogmatic theorist.

But we have not come to the end of this matter yet. Whilst M. La Place *proved* that no change greater than 1° 21′ could occur in the extent of the arctic circle, *because* the plane of the ecliptic could not vary more than that amount, and every follow-my-leader theorist bowed to the supposed infallibility and minute accuracy of this conclusion, we now are informed by another French theorist, M. Leverrier, that M. La Place committed an error; it is now stated that 4° 52′, not 1° 21′, is the amount that the plane of the ecliptic can vary.

Will these theorists kindly state the exact course which the earth's axis traces on the sphere of the heavens, and prove their knowledge by calculations?

CHAPTER IX.

THE SO-CALLED PROPER MOTION OF THE FIXED STARS.

THERE are few subjects which have attracted the attention of observers in modern times more completely than that termed somewhat peculiarly "the proper motion of the fixed stars."

The manner in which a catalogue of stars for any future date, and for the use of surveyors, navigators, etc., is formed is as follows:—

By the aid of a transit instrument placed in the plane of the meridian, the *zenith* distance of a star is measured, as often as it can be seen, on the meridian. The latitude of the observatory being known, it is also known what is the meridian zenith distance of the equator, and the zenith distance of the pole. For example, if the latitude of the observatory were 51° 28' north, the meridian zenith distance of the equator produced to the heavens would be 51° 28', and the zenith distance of the pole of the heavens would be 38° 32', the pole being north, the equinoctial (or equator produced) being south, of the zenith.

When, then, the meridian zenith distance of a star is measured with the transit instrument, the distance of this star from the equinoctial can be at once assigned, and this distance is the star's declination. For example, suppose a star were found to have a meridian zenith distance south of 10° 28', the latitude of the observatory being 51° 28', then

I

this star would be 41° north of the equinoctial, and would be given 41° north declination. If a star were to pass the meridian 10° north of the zenith, then this star would be given a north declination of 51° 28′ + 10° = 61° 28′ north. Year after year similar observations are repeated in order to obtain the rate of change annually in the declination or polar distance of a star. Thus, for example, suppose the declination of a star deduced from its meridian zenith distance were measured on the night of January 1, during several successive years, and were found as follows:—

January 1, 1850	..	..	..	..	30° 10′ 10·64″
,, 1851	..	..	..	..	30° 10′ 0·64″
,, 1852	..	..	..	..	30° 9′ 50·63″
,, 1853	..	..	..	..	30° 9′ 40·61″

The annual rate would be given to this star of −10″, with an increase in the rate of about two-hundredths of a second per year. It would be thus possible to fudge the declination of this star for four or five years in advance with very fair accuracy, without knowing in the least why the rate varied from 10″ to 10·01″, and then went to 10·02″.

For example, we could predict the declination of this star for January 1, 1854, by subtracting 10·03″ from the declination found for 1853, and we should thus assign the star a declination for January 1, 1854, of 30° 9′ 30·58″. We might, for January 1, 1855, subtract 10·06″ from that assigned for 1854, and so on.

The reader who will carefully consider the means adopted to obtain the declination of a star, and the annual rate of change in this declination, will perceive that the results are derived from instrumental observation only, and are not obtained in consequence of any knowledge of those mechanical movements in the earth which produce the change. The declination of the star is found by measuring its meridian zenith distance with the transit instrument.

SO-CALLED PROPER MOTION OF THE FIXED STARS. 115

The annual variation in the declination is found by comparing the declination found during successive years, due allowance being made for the nutation already referred to. If the earth were a fixed body having no daily rotation, but the whole sphere of the heavens revolved every twenty-four hours, the declination of a star, and the annual variation in this declination, could be obtained by exactly the same proceedings as are now adopted, and with the same ease and approximate accuracy.

Thus by observation only we may obtain a fairly accurate knowledge of the changes which take place annually in connection with various celestial bodies, although we may not have the slightest idea of the true cause of these changes.

As an example, a case may be mentioned which occurred more than two thousand years ago. Hipparchus, one of the ancient observers, determined by observation the longitude of the star Spica Virginis, and he found this star was 8° from the autumnal equinox. He found that one hundred and seventy years previously, two observers, viz. Timocharis and Aristyllus, had found the longitude of this same star 6° from the autumnal equinox. By the comparison of these longitudes, Hipparchus concluded that there was a precession of the equinoctial point of 2° in one hundred and seventy years.

Hipparchus did not know that the earth rotated on its axis during each twenty-four hours, or that it revolved round the sun during each year. He did not know that the earth's axis changed its direction, and thus produced this difference in longitude which he had found by observation. He imagined that the earth was stationary. Yet had his observations, or those of his predecessors, been a little more accurate, he might, by merely adding a certain amount annually to the then longitude of a star, have predicted with fair accuracy what would be the longitude of this star at some future period.

He would, by this rule-of-thumb method of adding a certain quantity annually, have given the longitude of a star for a future date by exactly the same method that the modern observer predicts the declination of a star for a future date, by adding or subtracting a given quantity for each year.

It is by no means difficult to become acquainted with various changes which occur, and which can be discovered by observation, but it requires some amount of thought and reason to discover the cause of these changes. Even the ploughman is aware that the days in summer are much longer than they are in winter—he knows this from observation; but he is not aware that the cause of this change is due to the fact that the earth's axis is at present inclined about 66° 33' to the plane of its orbit. That, if the earth's axis were at right angles to the plane of its orbit, the days during the year would be always of the same length, would be a truth which the ploughman could not comprehend.

It is, therefore, a very different thing to know from observation only that certain changes occur, and to know the causes which produce these changes.

The method of finding the declination or polar distance of a star by observation has now been described, and the reader must bear in mind that when the declination has been obtained, the polar distance is known, the declination added to the polar distance being always 90°.

It will be remembered that the declination of a star is deduced from its meridian zenith distance, measured by aid of the transit instrument, and the annual rate of change in the star's declination is obtained by the comparison of stars' assigned declinations from year to year.

The question of the so-termed "proper motion" of the stars may now be dealt with.

To those persons unacquainted with technical astronomy, the term "proper motion" of the stars might be understood to mean that a group of stars seen in the heavens altered their form or shape relative to each other. For example, suppose three stars not far from each other formed at a distant period an exact equilateral triangle, but that in the present day they formed an isosceles triangle; if such a change had occurred, it would be a proof that these stars had altered their relative position as regards each other, and had, therefore, some independent motion of their own.

It is not from such palpable changes that the conclusion has been arrived at that the stars have an independent movement of their own, but on account of their changes in *right ascension* and *declination* being different from what it is assumed they ought to be, in accordance with the accepted theory of the movement of the pole of the heavens.

It is quite possible, and even probable, that the stars have some independent movement among themselves, but the greatest caution is requisite, before we attribute to any stars such a motion, merely because their right ascension or declination changes in a manner not in accordance with the present accepted theories.

In the first place, we must bear in mind that the declination of a star is deduced from its observed meridian zenith distance, and its right ascension is obtained by its meridian transit.

The manner in which the zenith is affected by the second rotation of the earth, as shown in Chapter III., is very varied, and is dependent not only in amount, but in direction, upon the distance of this zenith from the pole of the second axis of rotation. Two localities on the same meridian of terrestrial longitude may have their zeniths affected in quite different manners by the second rotation, one zenith being carried directly towards the pole, the other being carried obliquely towards the pole. The following diagram will give some idea of these varied and important changes; and as these changes have hitherto been entirely unknown to astronomers, the reader's attention is particularly called to this subject.

P (Fig. 43) represents the pole of daily rotation on the sphere of the heavens; B and x, the zenith of two localities on earth, having the same terrestrial longitude, but differing in latitude by the arc x B.

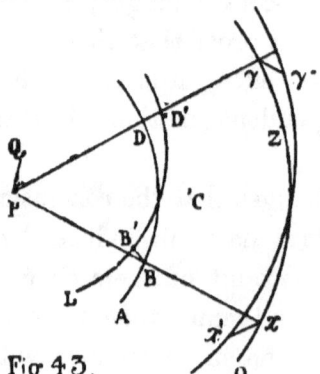

Fig 43.

B D is the arc traced on the sphere of the heavens, round P as a centre, by the zenith B during a portion of the *daily* rotation of the earth; $x \gamma$ is the arc traced on the sphere of the heavens, round P as a centre, by the zenith x during the same portion of the daily rotation as that which carried B to D.

During one daily rotation of the earth, the pole of daily rotation may be taken as a fixed point in the heavens.

We will now examine the effects which will be produced during any given time by the so-termed change in direction of the earth's axis, which movement really is a second rotation of the earth, and the point C represents on the sphere of the heavens the pole of the second axis of rotation.

The pole P is carried to Q, round C as a centre, by the second rotation, the arc P Q being taken of any length,

SO-CALLED PROPER MOTION OF THE FIXED STARS. 119

as an illustration of the laws affecting this movement of the earth.

At the date when the pole has reached Q, we take again a portion of the daily rotation of the two zeniths before referred to.

No change whatever in the *direction* of the axis of a rotating sphere can cause any change in distance between the pole of this sphere and of each zenith. Consequently we take Q L = P A and Q N = P O, and the arc L D' will be the arc now traced by the zenith which formerly traced the arc A B D during a portion of the daily rotation.

We now come to a most important item in connection with the changes of zenith—important because it has hitherto never been examined by geometricians.

The pole P is carried to Q by the second rotation of the earth, and the zenith B will be carried to B', the arc B B' being a portion of a small circle of the sphere having C for its centre.

In like manner, the zenith at D will be carried to D' by the same movement that has carried the pole from P to Q and the zenith B to B'.

The zenith which was at x will be carried by the second rotation to x', the arc $x\ x'$ being a portion of a small circle of the sphere the centre of which is C. Thus whilst the pole is carried from P to Q by the second rotation, and the zenith B is carried to B', the zenith x is carried to x' by the same movement of the earth.

The zenith γ will be carried to γ', round C as a centre, and the whole effect of a portion of the second rotation as affecting the pole and these zeniths is, that whilst P is carried to Q, B is carried to B', D to D', x to x', and γ to γ'.

The meridian of right ascension which, when the pole

was at P, would pass through B and x, would now, when the pole was at Q, be represented by a great circle of the sphere passing from Q through B' and x.'

The zenith B is carried by the second rotation to B', or nearly north, taking P or Q as the north pole of the heavens; but the zenith x is carried to x', an arc which is oblique as regards the North Pole. Consequently the zenith x', in order to reach its former meridian, of which P x is a part, has to be carried by the *daily* rotation over a small arc; whereas the zenith B, which has been carried to B' by the second rotation, has by this movement passed the meridian, viz. P x, on which it was formerly situated.

These varied changes, both in direction and in amount, over which various zeniths are carried by the second rotation, and the manner in which the meridian appears to change in consequence, afford some of the most beautiful yet simple problems ever submitted in connection with geometrical astronomy.

To those persons unacquainted with the second rotation of the earth, the most mysterious supposed movements of the stars will take place near the pole of the axis of second rotation, and also where the distance of a star from the pole of second rotation is greater than the distance of this same star from the pole of daily rotation.

This one branch of the second rotation is so varied in its effects, that many pages would be required to describe even a portion of these. A geometrician, however, will readily comprehend the principle, and can work out the details for himself, should he care to do so.

From even the brief description which has already been given as to the varied changes produced on a zenith and meridian by the second rotation, and varying very much according to where the zenith may be situated in relation to the pole of second rotation, we see that it is a somewhat hasty assumption to make, that because the meridian zenith distance of certain stars varies in an apparently irregular manner, therefore these stars have a "proper motion" of their own. We ought first to examine whether it may not be a proper motion in our zeniths and meridians, and not in the stars.

Another source of error in connection with the asserted proper motion of the stars, is due to the fact that the true radius of the circle which the earth's axis describes on the sphere of the heavens has hitherto been incorrectly estimated.

It is a geometrical law that the *relative* right ascension of any two stars on the circumference of the circle traced by the pole of the heavens will never vary. If, as has been the case, it were assumed that the radius of this circle were 23° 28′, then any two stars on the circumference of this circle would never vary their relative right ascensions. If by repeated observations it were found that the relative right ascension of these stars did vary, theorists would assert that one, or both of these stars, had a proper motion in right ascension.

If, however, the radius of the circle traced by the earth's axis happen not to be 23° 28′, but of any other value, then the variation in relative right ascension of these two stars does not prove that they have any "proper motion," but it proves that the circle assumed to be traced by the earth's axis has been incorrectly estimated.

To assume that a certain mechanical movement occurs in an instrument, and then to assert that all the apparent movements of distant objects which contradict this assumption are due to eccentric and varied movements of the distant objects, cannot be called a scientific method of investigation. It is, in fact, inverting the order in which inquiry should guide us to truth, inasmuch as a theory is first invented, and then facts are denied, when these contradict the assumed truth of the theory.

It is at present assumed that the radius of the circle which the earth's axis traces is 23° 28′, whereas the most exact calculations prove that the radius of this circle is 29° 25′ 47″. It follows, therefore, that there will be a point in the heavens where the circumference of these two circles are separated by the greatest distance, and at this point the greatest difference between the theoretical and actual position and changes of stars will be found. It may be interesting to note where this point in the heavens is situated, and the following diagram will show it:—

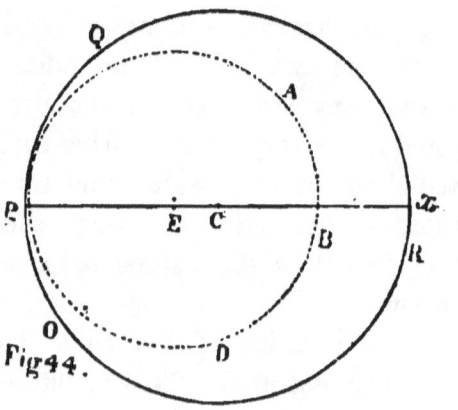

Fig. 44.

The point C is the centre of the circle which the pole traces in consequence of the second rotation of the earth. This circle is represented by O P Q x R, the radius being 29° 25′ 47″.

E is the centre of the circle A B D, which the earth's axis has been supposed to trace in consequence of "a conical movement," the radius E B being 23° 28′.

P is the position of the pole of the heavens at the date 1887. The arc P x will represent a portion of a meridian of eighteen hours right ascension.

At the point R the circumference of the larger circle will be most distant from the circumference of the smaller circle.

In what part of the heavens is this point R situated; and where, consequently, will the greatest differences be found between the theoretical and actual position of the stars?

First, as regards the polar distance of R. The radius C P = 29° 25' 47"; the radius C R = 29° 25' 47". The distance P R will be nearly equal to twice 29° 25' 47"— that is, to about 58° 51', which is about the polar distance of the point R. The point x has eighteen hours, or 270°, of right ascension; and the point R is a few degrees short of 270°—how many can be easily calculated, but, in round numbers, the right ascension of R would be about 265°.

A point in the heavens, having a north polar distance of about 58° 51', and a right ascension of 265°, is situated in the constellation Herculis, and is that point in the heavens where the true circle described by the pole of the heavens is at the greatest distance from the theoretical circle whose radius is assumed to be only 23° 28'.

It is singular that a point in the constellation Herculis almost coincident with the point mentioned above, has during many years attracted the attention of several theorists, in consequence of the stars in that region presenting the largest amount of difference between their theoretical and actual positions. Below will be seen the positions as assigned—about 58° 51' north polar distance, 265° right ascension, by comparing the results of the curve traced by the second rotation with the imaginary curve, with radius 23° 28'.

From the discordances found in the positions of stars, the point at which this discordance was greatest has been located as follows:—

North polar distance.	Right ascension.	By
59° 2'	261° 11'	Argelander.
62° 24'	261° 22'	Struve.
75° 34'	252° 53'	Luhndal.
55° 37'	260° 1'	Galloway.

That observers should have discovered that there was, about this point in the heavens, a marked difference between the theoretical and actual position of stars might be expected, considering that near that point there was a difference of 12° between the real and imaginary circle traced by the earth's axis. Few persons, however unacquainted with the tendency too often manifested to invent wonderful theories to explain simple geometrical laws, would have expected the conclusion arrived at by theorists to explain these discordances.

The theory that was put forward, at once accepted as true, that was copied by writer after writer, and asserted to be a fact, was, that the whole solar system was rushing towards this point in the constellation Herculis at the rate of one hundred and fifty-four million one hundred and eighty-five thousand miles per year (154,185,000 miles), and consequently it was merely a matter of time as to when a grand crash would occur, and the solar system be thereby destroyed.

Perhaps one of the most interesting items in connection with the history of astronomy, is to note the manner in which the general public and the tyro in the science, at once accept and repeat the most baseless theories, whilst they reject as absurd, conclusions which can be proved by exact geometry. No matter how absurd were the arguments employed to prove that the earth must be a flat surface, and could not be a globe, or that the earth must be a stationary body and could not rotate, yet these arguments were at once accepted as grand truths, whilst the real truth on these matters was considered a subject for

SO-CALLED PROPER MOTION OF THE FIXED STARS. 125

ridicule only. Had this mental feebleness not existed, the theory of the earth being stationary, could not have held in a state of slavery the minds of the scientific world during upwards of thirteen hundred years.

It is a fact that, at the point in the heavens in the constellation Herculis, there will be some marked changes in the positions of stars, but this is mainly due to the second rotation of the earth. Yet these changes never caused the slightest suspicion that the assumed circle traced by the pole might be incorrect.

Had it not been proved that the earth's axis traced a circle round the pole of the ecliptic as a centre, from which it never varied its distance? Was it not known that the joint action of the sun and moon on the earth's protuberant equator caused "a conical movement of the earth's axis"? To even question these theories was considered a proof of ignorance. Yet on so small a foundation as that great changes occurred in the observed right ascensions of stars near the constellation Herculis, the theory was invented that the whole solar system was rushing in that direction at an enormous rate, and that a grand crash must ensue.

To question this theory was asserted to be a proof of one's ignorance of the grand discoveries of astronomy.

It is always an unpleasant, and at the same time a most unpopular proceeding, to call attention to the errors which have been committed by those individuals, who have been long looked up to with veneration as authorities, and whose assertions or theories have been received with the greatest obedience. If, however, we desire to reach that which is true, we must overcome this unreasoning and blind submission to mere authority, otherwise we shall be committing the same errors as did the ancients, who denied that the earth could rotate, merely because all the great authorities during more than thirteen hundred years had asserted that it was stationary, and could not rotate.

There is probably no book, article, or paper written on the subject of the "proper motion" of the fixed stars which has tended so largely to promulgate error and mislead the mere follower as a paper by the late Professor F. Baily, in vol. v., "Memoirs of the Royal Astronomical Society," on the supposed method of finding the proper motion of the stars.

So unquestioned was the assumed accuracy of the method put forward in this paper, that the gold medal of the Society was presented to its author. Scores of followers, accepting as perfection the method suggested, set to work to make calculations as to the proper motions of stars, and hundreds of pages were filled with lists of assumed "proper motions," all based on the method of calculation recommended by this authority.

In order that the elementary error contained in the paper referred to may be perceived, the geometrical laws relative to a curve and a point must be studied.

Suppose A B C D (Fig. 45) a curve of any description; but, for the sake of illustration, it will be taken as part of the circumference of a circle, the centre of which is y.

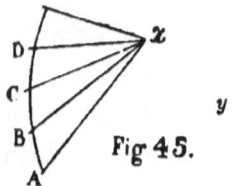

Fig 45.

Take any point, x, as a point of reference, and this point might be a star.

Take four points, A, B, C, D, equidistant from each other, and situated in the circumference referred to.

The distances A x, B x, C x, and D x, would represent the distances of x from four points in the circumference.

Suppose a body or a point of reference moved along the curve A D. This point would decrease its distance from x whilst moving from A to D. But the *rate* at which it decreased this distance would be a very variable rate, and in no two successive years would the rate be the same. Not only would the rate decrease, but it would *not decrease uniformly;* consequently, if we found the rate when the point was at A, and the rate when the point was at C, we should not obtain the rate when the point was at B by taking the mean of the other two rates.

This law holds good for all cases where a curve approaches or moves away from a point, and ought to be well known to every person acquainted with even the elementary laws of geometry.

The manner in which the sines, tangents, and secants of an angle vary gives another illustration of this law.

For example, the sine of 30° is 0·5000000. The sine of 34° is 0·5591929. If we took the mean of these two sines, we should obtain 0·52959645. But this is not the sine of 32°. The sine of 32° is 0·5299193.

Again, the sine of 30° increases for each minute of angle 2519. The sine of 34° increases for each minute of angle 2411. The mean of these rates is 2465, but this does not give us the rate of increase for each minute at 32°.

Small as the differences may be under some conditions, such as when the curve is moving directly away from or towards a point, yet it is a geometrical law that there must be an unequal variation in the rate at which a curve varies its distance from a fixed point.

Hence to compare the "rate" at one date with the "rate" found at another date, and to take the mean of this rate as a guide to the distance of a point from a curve, is a certain source of error.

The method proposed in the paper referred to is as follows: "Thus if P denote the place of a star in Piazzi's Catalogue (either in right ascension or declination), and B the place of the same star in Bradley's Catalogue, p the annual precession in 1800, π the annual precession in 1755, the annual proper motion μ' of a star for the first period (viz. 1755 to 1800) will be

$$\mu' = \frac{P-B}{45} - \frac{p+\pi}{2}$$

"And it is in this manner," says Professor Baily, "that I have deduced the annual proper motion of the stars in the list subjoined to this memoir."

With all due deference to this authority and his numerous followers, the laws of geometry prove that he will not obtain the proper motion of the stars by the method he has recommended. That which he will obtain is the difference between the mean rate at the middle period and the whole amount of the rate for the whole period, divided by the number of years between the two periods.

It is most remarkable that so palpable an error in elementary geometry as that contained in the paper referred to should have escaped the notice of the Council of the Royal Astronomical Society. It is equally as remarkable that the gold medal of the Society should have been given for the paper, and that followers by the score have copied this error and have framed tables of so-called "proper motion" which have no foundation in fact.

The error, however, is of a similar kind to that made by more modern theorists, who subtract a *constant* quantity annually in order to obtain the mean obliquity. In both cases the geometrical law, that a curve cannot approach a point at a uniform rate, and that this rate does not vary uniformly, has been overlooked.

By the method proposed by Professor Baily, nearly every star in the heavens would have assigned to it a "proper motion." Let us take some examples.

In a catalogue of stars for January 1, 1780, the north polar distance of the star β Draconis was found to be

37° 31′ 47″, and its annual rate of increase in polar distance was found to be 3·06″ annually.

In the Nautical Almanac for 1887, the mean north polar distance of this star is given as 37° 36′ 52·83″, and its annual increase in north polar distance 2·80″ annually.

Applying the method considered sound by Professor Baily and others, and substituting the items above, we should obtain the following results:—

$$\frac{37° \ 36' \ 52·83'' - 37° \ 31' \ 47''}{107} - \frac{3·06'' + 2·80''}{2} = \text{proper motion.}$$

That is, 2·93″ − 2·858″ = 0·072″ annually.

Consequently, it would be assumed that this star had a proper motion of 7·2″ in one hundred years. The fact really being that the rate 3·06″ does not change to 2·80″ uniformly, but has a "second difference," just in the same manner as the sine of an angle does not vary uniformly.

The annual rate at which a star is supposed to increase or decrease its polar distance has hitherto been obtained by observation, and is consequently liable to be affected by all those causes of error to which reference has been made. To attempt to deduce from such uncertain data definite conclusions as to the proper motion of stars, when in addition the method proposed is geometrically incorrect, is a somewhat loose manner of proceeding, and must lead to errors and false conclusions. When, also, the curve traced by the pole of the heavens, which is the cause of the change in polar distance of the stars, has never hitherto been accurately defined, it is evident that this subject has not been definitely disposed of. Had it been so disposed of, the present system of perpetual observation must be quite unnecessary, and can only be considered a useless piece of extravagance.

When we bring to bear on this question the knowledge of the second rotation of the earth, the confusion and inconsistencies vanish. The polar distance of a star and its annual rate can be calculated for 1000 years past or future as easily and more correctly than it can now be arrived at by any known system for even ten years.

The star β Draconis will be taken as an example.

The constants for this star are as follows:—

$$C \beta = 9° 17' 38''.$$
$$C P = 29° 25' 47''.$$
The angle P C β, January 1, 1887 = 148° 8' 0''.
Annual variation in the angle P C β = 40·9''.

Find the polar distance of this star for 1000 years in the past, viz. January 1, 887 A.D., and the annual rate of change in polar distance at that date. The rate of the second rotation being 40·9'' annually, the angle at C 1000 years ago would have been 40900'' = 11° 21' 40'' less than it was in 1887. Consequently, at 887 A.D. the angle at C was 148° 8' 0'' − 11° 21' 40'' = 136° 46' 20''.

The supplement of this angle is 43° 13' 40''.

We have now two sides and the included angle to find the third side P β, the polar distance of the star at 887 A.D. The whole of the working of this question is given below.

Log. cosine, 43° 13' 40'' = 9·8625110
Log. tangent, 9° 17' 38'' = 9·2139074
─────────
9·0764184 = log. tan. of first arc, 6° 47' 59''.

29° 25' 47''
6° 47' 59''
─────────
36° 13' 46'' = second arc.

Log. cosine, 9° 17' 38'' = 9·9942612
Log. cosine, 36° 13' 46'' = 9·9066887
─────────
19·9009499
− Log. cosine, 6° 47' 59'' = 9·9969344
─────────
9·9040155 = log. cos., 36° 42' 24''.

The polar distance of β Draconis, January 1, 887, was 36° 42' 24''.

We can now find the approximate annual rate of change in polar distance at that date, as follows :—

Substitute 1050 years for 1000 years, and the angle C for 837 A.D. will be 136° 12′ 15″, and its supplement, 43° 47′ 45″; proceed as in the last example, and calculate the polar distance of β Draconis for 837 A.D.

The calculation gives 36° 39′ 17″.

The difference for fifty years was therefore = 36° 42′ 24″ − 36° 39′ 17″ = 3′ 7″; that is, 187″ for fifty years, or at the rate of about 3·74″ annually at the date 887 A.D.

We can now test these results with the observations recorded in the Nautical Almanac for 1887.

The recorded north polar distance given in the Nautical Almanac for 1887 is—

$$\text{Calculated for 887} = \frac{37° 36′ 52·83″}{36° 42′ 24″}$$
$$54′ 24·83″ \text{ difference for 1000 years.}$$

54′ 24·83″ = 3268·83″ for 1000 years, or a mean rate of 3·268″ per year.

The annual rate given in the Nautical Almanac for 1887 for β Draconis is—

$$\begin{array}{r} 2·8″ \\ \text{Annual rate calculated for 887} = 3·74″ \\ \hline 2 \,|\, 6·54″ \\ \hline 3·27″ \end{array}$$

The difference between 3·27″ and 3·268″ is *two-thousandth of a second only;* but this difference even is not "proper motion" in the star, but is due to the fact that the annual rate does not and cannot vary uniformly.

It is a remarkable exhibition of the tenacity with which men blindly follow a routine system which is erroneous, to find a multitude of observers continuing to measure year after year the meridian zenith distances of stars, and imagining that the slight unaccountable differences which occur are due either to the proper motion of the stars, or to these stars having a parallax. How their zeniths and meridians are affected annually and semi-annually by "a conical movement of the earth's axis," is a question which they never seem to have imagined as of any importance. Hence we find pages of a scientific journal filled with the results of observations supposed to prove certain facts as regards distant stars, when these so-called facts have no foundation in truth. Thus errors are circulated far and wide, theories are based on these errors, and confusion reigns supreme. For men to base conclusions on the various meridian zenith distances of stars when they do not know how their zeniths or meridians move, and to refuse to examine how they do move, is a proceeding similar to that adopted by the followers of the system of Epicycles, who refused to examine the effects of the *daily* rotation of the earth.

CHAPTER X.

THE POLE-STAR.

PERHAPS one of the most important stars in the heavens, as regards astronomical observations, is the pole-star (α Ursæ Minoris).

This star, being at present very near the pole of the heavens, varies its altitude during each daily rotation of the earth only twice the amount of its polar distance, or less than $2\frac{1}{2}°$.

This star, consequently, can always be seen from such latitudes as those in which England and Europe are situated, and, if the night be clear, the meridian transit of this star can be observed each night during the year. It would occur, during summer, that the meridian transit, both above and below the pole, might take place during daylight. Yet, in spite of all difficulties, the meridian transit of this star, either above or below the pole, can be frequently observed.

It is sometimes claimed, as a proof of the important and valuable work carried on at an observatory, that more than one hundred observations have been made by observers of the pole-star during the year. Whether this work is as important as is imagined, will be left to the judgment of the reader who examines the facts in this chapter.

On bringing to bear on the pole-star the knowledge of

the second rotation, some interesting facts are revealed; and by aid of this the polar distance of this star will be calculated for the date 113 B.C.—that is, 2000 years from 1887 A.D.

The constants for the pole-star are as follows:—

C P (Fig. 46), from pole of second rotation to pole of daily rotation, 29° 25′ 47″.

C a, from pole of second rotation to Polaris, 29° 52′ 51″.

Angle a C P, January 1, 1887 = 2° 27′ 5″.

Annual variation in angle at C = 40·9″.

Fig 46. During 2000 years the angle at C would have been decreasing at the rate of 40·9″ per year; consequently this angle 2000 years in the past would have been 22° 43′ 20″ greater than it was on January 1, 1887. The angle C, at 113 B.C., was therefore 25° 10′ 25″.

With the two sides, C a = 29° 52′ 51″ and C P = 29° 25′ 47″, and the included angle at C = 25° 10′ 25″, the side P a can be calculated, and will be found 12° 23′ 16″, which was the polar distance of Polaris 113 B.C.

Fifty years previous, viz. at 163 B.C., the angle at C was 25° 44′ 30″, and, calculating as before, the polar distance of Polaris for 163 B.C. was 12° 39′ 46″.

Between 163 B.C. and 113 B.C. the polar distance of Polaris decreased 16′ 30,″ being at the rate of 19·8″ per year, between those periods.

Comparing these items with those in the Nautical Almanac for 1887, we obtain the following:—

Polar distance Polaris 1887 = 1° 17′ 38·2″
Polar distance Polaris 2000 years previous = 12° 23′ 16″
Difference = 11° 5′ 37·8″

The difference 39937·8″ divided by 2000 gives 19·967″ for the mean rate between these two periods, according to the difference in polar distance at the two dates.

THE POLE-STAR. 135

The rate of decrease in polar distance 113 B.C. = 19·8"
The rate of decrease 1887 A.D. in Nautical Almanack = 18·923"
$$2\overline{)38\cdot 723"}$$
Mean 19·361"

We have here an excellent opportunity of testing the method recommended by Professor Baily for finding the proper motion of stars. We find 19·967" the rate of change according to the difference in polar distance of the star Polaris at the two dates, but we find 19·361" as the mean of the rates at the two periods.

According to the system recommended by this authority, the difference between these two items, viz. 0·606", would be assigned as proper motion to the pole-star annually.

Here is an admirable opportunity for the theorist to step in and to invent some wonderful theory as to the cause of this supposed proper motion in the pole-star.

Every person, of course, knows that the whole solar system is rushing towards the constellation Herculis; this movement has been accurately proved by authorities. Now, however, something equally as wonderful can be invented.

As the pole-star has such a large proper motion, the solar system is probably also rushing towards the pole-star, and a grand collision must occur between the sun and the pole-star, with results fearful to contemplate.

This theory is recommended to those gentlemen who lecture to audiences, with a view to astonishing these, for it is not more baseless and untrue than many statements on astronomy which, when made, are loudly cheered by the wonder-loving listeners.

The pole-star has no "proper motion."

The system practised in order to find the proper motion is incorrect.

The difference, viz. 0·606" per year, found by taking

the mean rate by the two systems, is an additional proof, were one necessary, of the true movement of the pole, and of the accuracy of geometry.

We have now before us a problem worthy the attention of geometricians and astronomers, and of a very different class from that of merely adding or subtracting a quantity found by observation, in order to give the polar distance of a star for half a dozen years in advance. This is, to show why there is this difference of 0·606″ as found above, and to prove that this is not " proper motion " in the pole-star.

The course which the pole of the heavens traces on the sphere of the heavens, in consequence of the second rotation of the earth, is a circle, having for its centre a point 29° 25′ 47″ from the pole of daily rotation. This rotation occurs at the rate of 40·9″ annually.

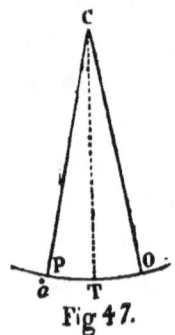

Fig 47.

At the date 1887 A.D. the pole was at P (Fig. 47); at the date 113 B.C. the pole was at O; the pole-star being a fixed point at a, and C, the centre of the circle traced by the pole, being a fixed point at C.

From 113 B.C. to 1887 A.D. the pole has been carried along the curve O P.

When the pole was at O, the annual rate of decrease of the pole from a was 19·8″. When the pole was at P, the annual rate of decrease of P from a was only 18·923″, and the mean of these rates gives only 19·361″. Whereas when the less polar distance is taken from the greater, and the remainder divided by 2000 (the number of years between the dates), we find that the pole must have approached the pole-star at a mean rate of 19·967″ per year.

It would be very easy, but very erroneous, to put this apparent inconsistency down to " proper motion ; " but the true cause is so very simple, and proves so accurately the

THE POLE-STAR. 137

course of the pole, that it is quite unnecessary to invent any fantastic theories to account for the facts.

The truth is, that the pole, when at O, was not moving directly towards the pole-star, and when at P the pole was not moving directly towards the pole-star, but somewhere between these two dates the pole *was* moving directly towards the pole-star, and was therefore decreasing its distance annually from the star, at a rate greater than that at which it decreased it either at O or P.

When was the pole moving directly towards the pole-star, is now the question for solution.

The problem is a simple one. The pole was moving directly towards the pole-star when an arc joining the pole-star and the pole was at right angles to the arc joining the pole of second rotation with the pole of daily rotation, and to discover the date of this event presents no difficulties.

We have merely to investigate the case of a right-angled spherical triangle, as follows:—

C a (Fig. 48) is the hypotenuse = 29° 52′ 51″; from the pole of second rotation to the pole-star P, C P = 29° 25′ 47″.

The angle at P is a right angle when the pole is moving directly towards a.

The value of the angle at C has now to be calculated for these conditions, as follows:—

Fig 48.

```
Log. tan., 29° 25′ 47″ = 9·7513982
Log. cotan., 29° 52′ 51″ = 0·2406491
                          ─────────
                          9·9920473 = Log. cos. < C = 10° 55′ 53″.
```

At the date 1887 the angle at C was 2° 27′ 5″, which subtracted from the above angle, leaves 8° 28′ 48″ for the angle of second rotation between 1887 and the date when the pole was moving directly towards the pole-star.

8° 28′ 48″ = 30528″, which, at the rate of 40·9″ per year, occupied 746·4 years.

746·4, taken from 1887, gives the date 1140·6 A.D. for the period when the pole of daily rotation was being carried directly towards the pole-star, and when, consequently, the annual decrease in polar distance of this star was greatest.

The rate at which the pole, at the date 1140·6 A.D., approached the pole-star was 40·9″ × sine of 29° 25′ 47″ = 20·09″ per year.

From the date 113 B.C. to the date 1140·6 A.D., the rate at which the pole increased its rate of approach to the pole-star was from 19·8″ at 113 B.C., to 20·09″ at 1140·6 A.D. From 1140·6 A.D. up to the present time the rate has decreased, though by no means uniformly, until it has reached the rate at present found by observation of about 18·923″ annually.

From the present date the annual rate will decrease rapidly, until at the date 2102·76 A.D. the pole will not vary its distance from the pole-star for that year more than 1″. The polar distance of the pole-star will then be 27′ 4″, the nearest approach of the pole to the pole-star. After the date 2102·76 A.D. the pole will increase its distance from the pole-star *nearly* in the same manner as it decreased this distance.

We trust that the reader will now be able to work out such simple geometrical problems as the mean polar distance of the pole-star, for thousands of years, without any reference to observations.

It may, therefore, fairly admit of question, whether it is a proof that very important work is carried on at official observatories, when it is stated that one hundred observations at least have been taken each year of the pole-star alone.

It will occupy observers nearly 215 years to find out

that which a geometrician acquainted with the second rotation of the earth can calculate during ten minutes.

Let us note, however, the wonderful state of confusion into which theorists and the unreasoning followers of authority would have placed themselves by following the advice of the supposed best method of obtaining the proper motion of the fixed stars.

At the date 113 B.C. the rate was 19·8″, at 1887 it was 18·923″, giving a mean of 19·361″. Yet when the polar distance at 113 B.C. is compared with the polar distance at 1887 we obtain 19·967″. Hence, according to a great authority, the difference, viz. 0·606″, is to be put down as the annual proper motion of the star. This assumed annual "proper motion" is imagined to be arrived at by so accurate a method that theories can be based on it, and promulgated as though they were facts correctly proved by geometry. Whereas the very first step is erroneous, and the conclusions based thereon are as false as though we founded theories on the assumption of the earth being a flat surface.

It is certainly very remarkable that in the present day, when we have so much talk about competitive examinations, and the advantages of a mathematical and geometrical training, that we should yet find men, claiming to be scientific, submissively copying the elementary errors of those predecessors who are regarded as authorities. It is equally as curious to find also that there are persons who are employed to write in papers, professing to teach science to the public, who seem to imagine that when, like parrots, they can say that "the joint action of the sun and moon on the earth's protuberant equator makes a shift (*sic*) in the earth's axis," that they have uttered something very profound.

Such writers must imagine that the intellectual capacity of their readers is of a very low order, or they would not presume to string together such feeble nonsense.

From such proceedings, however, we find a probable solution of the mystery as to how it was possible that, during two thousand years after the daily rotation of the earth was suggested, this fact was sneered at and rejected by the so-called scientific authorities of the civilized world, whilst the dogmatic theory of Epicycles and a stationary earth was stated to be such a grand and difficult science, that those persons who questioned its accuracy were asserted to be ignorant and stupid.

The Star ε Ursæ Minoris.

This star will serve as a second example of the law relative to the variation in rate at which the pole of the heavens varies its distance from a star. The following diagram (Fig. 49) shows the relative position of the star, the pole of daily rotation, and the pole of second rotation at the date January 1, 1887.

C P = 29° 25' 47"; that is, from C, the pole of second rotation, to P, the pole of daily rotation.

C ε = 22° 1' 44·7", from C to ε, the star.

P ε for January 1, 1887 = 7° 46' 41·6". The angle P C ε January 1, 1887 = 5° 34' 14·9". The annual variation in the angle P C ε = 40·9".

When the pole P was, at some date in the past, situated at O, so that the star ε was on the arc joining O and C, the pole O would then be moving at right angles to the arc joining O and ε, and at this date there would for an instant be no increase or decrease in the polar distance of the star ε.

The polar distance of this star would then be C O − C ε = 29° 25' 47" − 22° 1' 44·7" = 7° 24' 2·3". At the date January 1, 1887, the polar distance of ε was found to be 7° 46' 41·5"; the difference is 0° 22' 39·2".

The date at which the pole was at O can be found by dividing 5° 34' 14·9" by 40·9", which gives 490·4 years from 1887, that is at 1397·6 A.D.

Consequently, during 490·6 years the polar distance of this star has increased 22' 39·2"; that is, if we took the rate as uniform, at a mean rate of 2·76" annually.

But at 1397·6 A.D. the annual rate was nothing, whilst in the Nautical Almanac for 1887 this rate is given as 5·390". Hence, to discover the supposed proper motion of this star, we should take the rate at 1397·6 A.D. from the rate at 1887 A.D., divide by 2, and compare this result, viz. 2·69", with 2·76" found above, and the difference 0·07" would be attributed to the proper motion of the star.

The error of this method is so palpable that it seems almost unnecessary to demonstrate it, yet reference to it cannot be avoided when we find that a council of learned gentlemen considered its discovery and the conclusions based thereon were so valuable as to deserve the gold medal of the society.

The two stars, Polaris and ε Ursæ Minoris, to which reference has been made, serve to show the principle on which the polar distance varies, and how the rate varies. The same principle, however, holds good for other stars, although, in a multitude of instances, to a less degree than it does to those stars which have been dealt with.

The method of finding the exact polar distance of a star for any date when we are acquainted with the second rotation of the earth, with the rate of this second rotation, and consequently with the true course of the pole, may possibly appear to superficial investigators a very easy and simple problem. There are, however, certain laws affecting this problem, and dependent on the present method of making observations, which must not be overlooked.

It must be borne in mind that observers in the present

day do not measure the polar distance of a star with their transit instrument. That which they measure is the meridian zenith distance of the star, and they deduce the declination or polar distance of the star from the measured meridian zenith distance.

The manner in which the zenith and meridian are affected by the second rotation of the earth becomes, therefore, a most important item for inquiry, and it will be found that it is one which will fully repay those geometricians who will take the trouble to investigate it.

CHAPTER XI.

THE NAUTILUS CURVE.

IN the preceding pages mention has frequently been made of the *mean* polar distance of a star, and the *mean* obliquity for some particular year. To the reader unacquainted with the details of astronomy, this term "mean" has no doubt been unintelligible.

Until about a hundred and fifty years ago, observations were made so imperfectly, that it was imagined that the pole of the heavens appeared to move uniformly during the year over its small arc of about 20·09″ annually.

Bradley, a careful observer, about a hundred and fifty years ago, used an instrument termed "a zenith sector," with which he could measure meridian zenith distances with great accuracy. By the aid of this instrument, Bradley found that the zenith distance of the star γ Draconis varied its zenith distance during the year in a manner that had not been accounted for by any theorist.

At first Bradley concluded that the cause of this change in the zenith distance of the star might be due to some movement of the earth, and hence of the zenith, which had hitherto escaped notice. He soon, however, gave up this idea, and stated that this change in zenith distance of the star must be due to the velocity of light and the movement of the earth.

This theory was at once accepted, and was termed

"aberration," a title by no means inappropriate if applied to something else, in addition to the stars.

The effect of this "aberration" is to cause the stars to vary their distance from the zenith during the year, in a manner which will be described; hence, when the mean polar distance of a star is referred to, the term indicates the polar distance of the star if it were unaffected by aberration, and also the small displacement produced in the pole by the "nutation," or small elliptical movement of the pole, which occupies slightly less than nineteen years.

In order to make clear the effect produced annually on the zenith distance of a star, and on its supposed polar distance, and also on its right ascension, the nautilus curve will be made use of.

The nautilus curve is constructed as follows:—

Take a nearly straight line, O P, of any convenient length, and divide this line into twelve equal parts, shown by the numbers 1, 2, 3, etc. This line is actually 20·09" of the arc of a circle the radius of which is 29° 25' 47".

With O as a centre set off the angle A O P = 30°, and with O as centre describe the arc P A.

From A 1 set off the angle A 1 B = 30°, and with 1 as a centre and radius 1 A describe the arc A B.

From B 2 set off 30°, and with 2 as a centre and radius 2 B describe the arc B C.

From C 3 set off 30°, and with C as a centre describe the arc C D.

Proceed in the same manner, setting off twelve angles of 30° each, amounting to 360°, and the nautilus curve as show in the diagram and defined by P, A, B, C, D E, etc. is constructed.

From P to A is termed January, A to B is termed February, B to C, March, and so on, as written in the diagram.

The distance from O to P must be taken as 20·09", and

a scale must be constructed in accordance therewith, as shown below the curve in the diagram. From this scale decimals of seconds can be measured.

The production of the line O P towards Q will be the

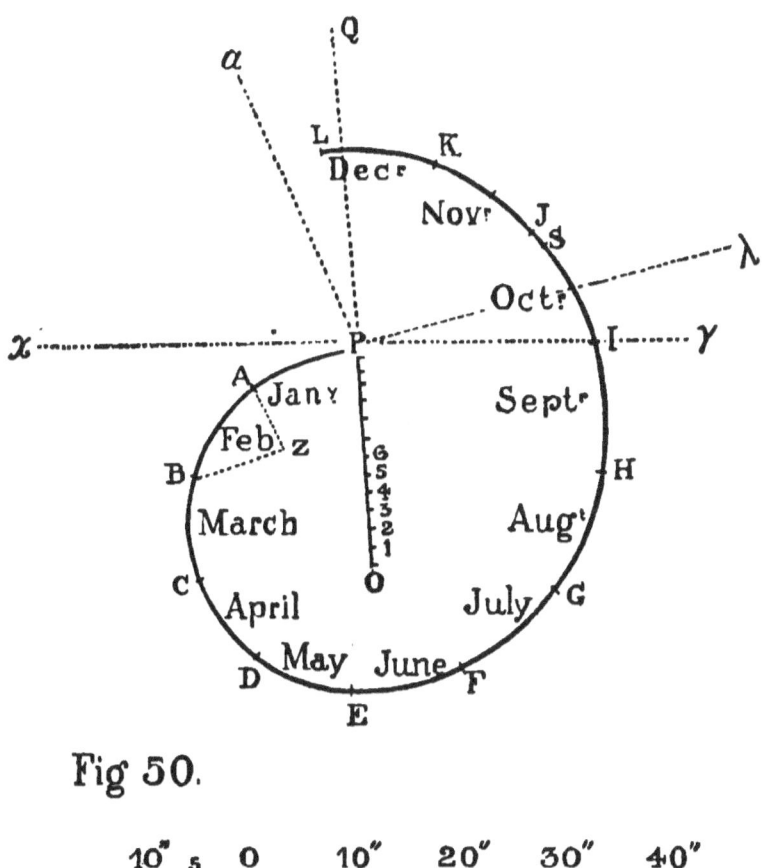

Fig 50.

direction in which stars are situated which have twenty-four hours of right ascension.

The direction P x, at right angles to P O, will be the direction in which stars are situated which have six hours right ascension.

The line P O produced will point to twelve hours right ascension, and P γ to eighteen hours right ascension.

This curve being very small, a star more than 1° from P will be so situated that lines or arcs drawn from the star to any part of the curve will be *nearly* parallel to each other. These lines will not be quite parallel, as they must form an angle at the star, as will be shown, but they are nearly parallel. The more distant a star is from P, the pole of the heavens, on January 1 of any given year, the more nearly will the lines drawn from such star to any part of the curve be parallel to one another.

The application of this curve will now be demonstrated.

The pole-star, α Ursæ Minoris, had on January 1, 1887, an *apparent* right ascension of 1h. 17m. 49·23s., and a polar distance (apparent) of 1° 17′ 24″, as stated to have been found by observation, and recorded in the Nautical Almanac for 1887.

1h. 17m. 49·23s. of right ascension converted into degrees and minutes will amount to about 19° 27′ 19″.

From P set off the angle Q P α = 19° 27′ 19″, and the line or arc P α will be the *direction* of the pole-star.

The distance of the pole-star from P on January 1, 1887, was found by observation to be 1° 17′ 24″, which distance expressed in seconds is 4664″. Hence if the pole-star's position were plotted on a diagram, on the same scale as that on which the nautilus curve is drawn, the pole-star would be in the direction of P α, and about thirty feet from P.

It will then be seen that any lines or arcs drawn from a point thirty feet distant, and to any part of the nautilus curve, would be *nearly* parallel to each other.

The observations that are now made in order to determine the polar distance of a star are in reality meridian zenith distances, the polar distance being deduced from the zenith distance. Any changes, therefore, which may occur in any zenith, or in any meridian, or in any part of a meridian, would be transferred, as it were, by the present system, to a supposed movement in the pole of the heavens, or in the star itself.

Let us now assume that these supposed movements are such as to cause the pole of the heavens to appear to move round the nautilus curve during a year, the course for each month being indicated on the diagram. The apparent effect as regards the pole-star's polar distance will now be examined, and the effects on its apparent right ascension will afterwards be dealt with.

The apparent movement of the pole from P to A during January will cause the curve thus traced to slightly decrease its distance from the pole-star. How much can be measured as follows:—

The curve being P A, join P A by a straight line A x P, and measure the greatest distance between the straight line and the curve. The greatest distance will be found at x, and by scale is $\frac{8}{10}$ of a second, and would occur about the middle of January.

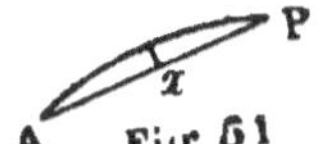

Fig. 51

On reference to the Nautical Almanac, 1887, page 310, it is recorded that the apparent polar distance of the pole-star found by observation, was as follows:—

January 1	1° 17′ 24″
January 13	1° 17′ 23·2″
January 21	1° 17′ 23·2″
January 31	1° 17′ 23·8″

Thus between January 1 and the middle of the month, we have exactly $\frac{8}{10}$ of a second decrease in the polar distance of Polaris, as indicated by the nautilus curve; whilst as the curve approaches A the polar distance again increases.

We can now examine the curve during the next month, February, viz. from A to B.

From an examination of the curve, it is evident that during February the polar distance of Polaris will increase, and the amount from the 1st to the 28th of February can be accurately ascertained as follows: From A draw A Z parallel to a P, and draw B Z at right angles to A Z, the distance A Z measured by scale will give the increase of polar distance of Polaris during February. This distance by scale will be found 4·4" or 4·5".

In the Nautical Almanac for 1887, the polar distance for Polaris is recorded as follows:—

February 1 .. .. 1° 17' 23·9"
February 14 .. .. 1° 17' 25·6"
February 28 .. .. 1° 17' 28·5"

During the first half of the month the decrease is only 1·7", during the second half it is 2·9", making a total of 4·6".

Why the increase in the early part of the month is less than it is during the second half, will be evident by examining the curve between A and B.

An examination of the next month, March (from B to C), gives the distance for the increase in polar distance—about 8·8" by scale.

In the Nautical Almanac for 1887, the polar distance of Polaris is recorded as follows:—

March 1 .. .. 1° 17' 28·7" } difference, 8·8
March 31 .. .. .. 1° 17' 37·5"

From an examination of the curve, it will be evident that during April the curve is moving more directly away from the pole-star than it is during either March or May; therefore between April 1 and May 1 there would be a larger increase in the polar distance than during thirty-one days in March. The amount by scale is about 9·5". The amount given in the Nautical Almanac is—

April 1 .. .. 1° 17' 37·8" } difference, 9·5"
May 1 .. .. 1° 17' 47·3"

THE NAUTILUS CURVE. 149

The next question to be investigated by aid of the curve is, at what date will the polar distance of Polaris be a constant? It will be a constant when the curve is moving at right angles to the arc joining the pole-star and the curve, a condition which will exist close to F in the curve.

In the Nautical Almanac, 1887, the polar distance of the pole-star is given as a constant, viz. 1° 17′ 55·4″, from the 23rd to the 30th of June. From June 30 the polar distance of the pole-star will decrease, as shown by the curve, the amount in each case being capable of measurement as already explained.

A very interesting fact will now be pointed out in connection with the nautilus curve, viz. that when this curve is moving directly towards the pole-star, the rate of decrease in polar distance will be at its greatest, and this will occur between I and J. But the distance between I and J is, by the construction of the curve, greater than is the distance between C and D. Hence the decrease for thirty days during October should be greater than the increase during thirty days in April.

Measuring on the curve from I to J, we obtain about 11·0″ for the decrease.

In the Nautical Almanac for 1887, the following is recorded:—

October 1 .. .. .. 1° 17′ 33·9″ } difference, 10·9″
October 30 .. .. .. 1° 17′ 23·0″ }

Whereas, during thirty days in April, the increase in polar distance is only 9·0″.

As the curve is carried round from J to K and on to P, the polar distance of the pole-star decreases, but not uniformly, the rate from day to day decreasing.

When the curve approaches J near the end of October there will be found a point S in the curve, where a line drawn from P at right angles to P a cuts the curve. The distance from this point S to the pole-star will be equal to the distance from P to the pole-star, the small distance P S, compared to the great distance P a and S a, causing P a and S a to be practically two sides of an isosceles triangle.

The distance J S will be found by scale one-ninth of J S. Consequently, at about October 28, when the curve had reached S, the polar distance of the pole-star would be almost of the same value as it was on January 1 of the same year.

In the Nautical Almanac for 1887, we find the polar distance of the pole-star recorded as follows:—

January 1   .. ..   1° 17′ 24″   } difference, $\frac{3}{10}$ of a second.
October 28   .. ..   1° 17′ 23·7′

When the curve has reached L the 360° of the nautilus curve has been completed, and the arc $\dot{P}$ L indicates the apparent change in the position of the pole from January 1 to January 1 of the year following.

In order to find how much nearer the point L is to the pole-star than the point P, we have merely to multiply the distance P L taken from the scale = 20·09″ by the cosine of the angle L P a, which angle is, for January 1, 19° 19′ 45″, and we obtain 18·94″ for the decrease in polar distance of the pole-star during the year 1887.

In the Nautical Almanac for 1887, the decrease is recorded as 18·92″, giving a difference of two-hundredths of a second.

The reader will now probably comprehend that the effects of what is termed "aberration" are considerable during a year, and produce great changes in the supposed polar distance of a star near the pole. The other effect produced by what is termed "nutation" is slight by comparison, and at present need not be referred to.

It will be evident that when an observer unacquainted

with the effect produced by aberration (viz. observers who made observations more than a hundred and fifty years in the past) states that he found the polar distance of the pole-star of a certain value, but does not state the day of the month, but merely gives the year, his records are not of much value, inasmuch as, if he made these observations at the end of December, they would differ fully 50″ from those made about the end of June of the same year.

The effect on the apparent right ascension of the pole-star by the nautilus curve will now be briefly described.

As the curve moves from P to A, the right ascension of the pole-star must decrease, and it will continue to decrease until a line or arc from the pole-star is a tangent to the curve. This, from an examination of the curve, it is evident, will take place near C in the early part of April.

In the Nautical Almanac for 1887, April 7 is the date at which the right ascension alters its rate from decreasing to increasing.

Again, it will be seen that about the middle of October, between I and J, a line from the pole-star to the curve will be again a tangent, at which date the right ascension of the star will vary from an increase to a decrease.

In the Nautical Almanac for 1887, the change is recorded as occurring between October 13 and 16.

Thus from January 1 until about April 7 the right ascension will increase; from April 7 till October 15 the right ascension will decrease.

These dates will vary slightly each year, inasmuch as O P is really a curve, being a portion of the circle described by the pole of daily rotation round the pole of the second rotation as a centre.

Another star, viz. λ Ursæ Minoris, will now be examined in connection with the nautilus curve.

This star on January 1, 1887, had a right ascension of

19h. 35m. 57s.; the direction of this star is plotted on the curve. It was distant by observation at that date 1° 2' 12" from the pole.

On examining the curve, it will be seen that the curve from P to A moves almost directly away from this star, the amount by scale being about 10".

In the Nautical Almanac for 1887, the polar distance of this star is given as follows:—

January 1 .. .. 1° 2' 12" } difference, 10·2"
January 31 .. .. 1° 2' 22·2"

The increase in polar distance will continue until a line from the star to the curve is at right angles to the curve, which will occur at about one-third of the distance between C and I, or about April 10.

The right ascension will slightly decrease through a portion of January; it will then increase.

A line drawn from the star through P, and produced, cuts the curve about one-third of the distance from A to B, viz. about Feburary 10, at which date the right ascension will be the same as it was when the curve was at P.

In the Nautical Almanac for 1887, page 310, we find the following:—

	h. m. s.
Right Ascension λ Ursæ Minoris, January 1 ..	19 35 57·87
,, ,, ,, February 10 ..	19 35 57·16

When a line from the star to the curve is a tangent to the curve, the right ascension will not vary during a day. From an examination of the curve, it will be seen that such will be the case just beyond F, or near the beginning of July.

In the Nautical Almanac for 1887, we find the following recorded for λ Ursæ Minoris:—

	h. m. s.
Right Ascension, June 30	19 37 52·53
,, July 1	19 37 52·62
,, July 2	19 37 52·63
,, July 3	19 37 52·53

When a line from the star to the curve cuts the curve at right angles, there will at that date be no change in the polar distance of the star during two or three days.

From an examination of the curve, it will be seen that this will occur about midway between I and J, or during October.

In the Nautical Almanac for 1887, the following are the recorded polar distances of this star :—

October 15	1° 1' 54·4"
October 16	1° 1' 54·4"
October 17	1° 1' 54·4"
October 18	1° 1' 54·3"
November 1	1° 1' 54·4"

When the curve has reached L, on December 31, the polar distance of λ Ursæ Minoris will have decreased since January 1, $20·09'' \times$ cosine of the angle $Q P \lambda = 20·09'' \times$ cosine $66° = 8·171''$, which will be the approximate rate of decrease in polar distance of this star for the year 1887.

In the Nautical Almanac for 1887, the rate given for this star's decrease in polar distance is $8·206''$, showing a difference of thirty-five-thousandths of a second.

The reader may take another star near the pole, viz. the star Cephei 51 Hev., the right ascension of which (apparent) on January, 1887, was 6h. 47m. 37·44s., and plot the direction of this star from P on the curve.

47m. 37·44s. correspond in round numbers to 11° 44' 21". Consequently, by making at P an angle x P 51 Hev. 11° 44' 21", we obtain the direction of this star.

Every detail as regards the annual changes in Cephei 51 Hev. can be read off the nautilus curve; the dates at which the right ascension will be a constant, at which the polar distance will be a constant, etc., can be obtained at a glance.

Any person who may not be provided with an observatory or an expensive transit instrument can, by aid of the nautilus curve, measure the changes which occur annually in a star *close* to the pole, and can understand the principal causes of those differences in the polar distance or right ascension of such stars, and described by professional observers as produced by "aberration."

Other stars near the pole may be selected and plotted on the diagram, such, for example, as δ Ursæ Minoris, λ Ursæ Minoris, and ε Ursæ Minoris, but an allowance has to be made on the annual effects when the star is at any great distance from the pole.

The nautilus curve for the south pole will be in the opposite direction to that for the north pole, so that the curve appears as though seen through a transparent paper. The shape of the curve will be the same as for the northern pole, but merely reversed. The reader can construct such a curve for himself on any scale he may select.

When a star is at a considerable distance from the pole of daily rotation, the annual changes as read by the nautilus curve will be the same in their general effect, though reduced in amount. Thus, for example, take the star γ Draconis, which has a right ascension of about 17h. 54m.

This star, it will be seen from an examination of the nautilus curve, will increase its polar distance rapidly during January, as the curve moves from P to A. It will increase its polar distance, but less rapidly, as the curve moves from A to B. At the end of March, viz. near C, a line from the star to the curve cuts the curve at right angles, at which date the polar distance will be a constant.

The following extract from the Nautical Almanac for 1887 gives the polar distance of γ Draconis on the first and other days of the month:—

January 1	38° 29′ 47·4″
January 31	38° 29′ 57·2″
February 10	38° 29′ 59·7″
March 2	38° 30′ 3·2″

THE NAUTILUS CURVE. 155

March 12	.. ..	38° 30' 4·0''
March 22	.. ..	38° 30' 4·2'' ⎫
April 1	.. ..	38° 30' 3·7'' ⎬ changes.
April 21	.. ..	38° 30' 0·9''

When a line from the star to the curve is a tangent to the curve, the right ascension of the star will not vary. An examination of the curve shows that this will occur about midway between E and F; that is, about the end of June.

In the Nautical Almanac for 1887, the following are the recorded right ascensions of γ Draconis:—

	h. m. s.
May 31	17 54 1·21
June 10	17 54 1·34
June 20	17 54 1·41 ⎫
June 30	17 54 1·42 ⎬ changes.
July 10	17 54 1·37
July 20	17 54 1·26

The right ascension will then decrease until about the middle of December, when a line from the star is again a tangent to the curve.

Whilst, however, the polar distance of the star λ Ursæ Minoris varies during the year as much as 41·3'', that of γ Draconis will vary only about 38·2''.

The nearer we approach the equator of second rotation, the less will the change be in polar distance of a star thus situated. Hence, if we project the curve on the plane of the equator of second rotation, we obtain a very close approximation to the effects produced as regards polar distance.

Consequently, a star with eighteen hours right ascension, and, say, 38° 40' of north declination, would be about 68° 5' from the equator of slow rotation; whereas a star with six hours right ascension and the same declination would be only 8° 15' from the equator of slow rotation. The star with eighteen hours right ascension would vary its polar distance considerably during the year; the star with six hours right ascension would vary it very slightly during the year. The two stars α Lyræ, with about 18h. 33m. right ascension, and Castor, with about 7h. 38m. right ascension, will serve as examples.

The reader who is desirous of investigating the nautilus curve may test its effects as regards the general results on the following stars, the direction of which he can draw on the diagram:—

Star.	Right ascension approximate.
	h. m.
θ Ursæ Majoris	9 25
α Ursæ Majoris	10 56
λ Draconis	11 24
γ Ursæ Majoris	11 48
α Draconis	14 1
β Ursæ Minoris	14 51
s Ursæ Minoris	15 48
η² Draconis	16 22
ε Ursæ Minoris	16 57
α Cephei	21 16
β² Cephei	21 27
γ Cephei	23 34

He will find, on reference to the nautilus curve, that the right ascension varies exactly in the manner that has been described, and that the polar distance also varies in the manner described, the amount of change in polar distance being decreased slightly as the star's place is nearer the equator of slow rotation.

A comparison can be made of the changes in right ascension, etc., as found in the nautilus curve, and those recorded in the Nautical Almanac for each star as found by observation.

The nautilus curve, although probably a novelty to the reader who is not acquainted with everything in the science of astronomy, ought to be thoroughly known to astronomers. About ten years ago, when I had tested the accuracy of this curve as a means of demonstrating the annual changes in right ascension and polar distance of important stars, I thought, as it was quite new, it might be of interest to astronomers. I therefore drew the curves for the northern and southern hemispheres, and, adding a full written description with numerous examples, I forwarded these to the Royal Astronomical Society.

Being aware that the pages of the monthly notices of the society were, to a great extent, filled with the recorded observations of various observers who did not seem aware how their zeniths or meridians were affected by that movement of the earth, vaguely defined as "a conical movement of the axis," I believed that attention might be directed to something novel, and certainly interesting.

I found, however, that I had made a mistake—that to submit for investigation an original subject was a most improper proceeding. Had I worked out the assumed proper motion of some hundred stars by means of the wonderful formula proposed by Professor Baily, my routine work would probably have been considered of some value, and would most likely have been printed and circulated; but the curve herein described was a novelty treated in a different manner.

I was informed that my paper had been received, and was placed among the records of the society—a polite way of describing "pigeon-holing" the document.

I would not for a moment question the soundness of the proceedings of those who considered it desirable that my paper should be suppressed. It is probably due to some feebleness of intellect on my part that I am puzzled to comprehend why the gold medal of a scientific society should be given to a gentleman for a paper containing, as a base, a most elementary error in geometry, whilst another paper giving an original curve, by aid of which the changes in stars can be read off, was suppressed.

If, some two thousand years ago, when the learned authorities all agreed that the earth was a flat surface, a geometrician had submitted to these gentlemen a paper giving such a proof of the spherical form of the earth as is contained in Chapter I. of this book, it is easy to predict the treatment which such a geometrical proof would have received. The audacity of any man who presumed to question the infallibility of the opinions of these ancient authorities must be put down at once, his paper suppressed, and his sound proofs treated as the learned Herodotus tells us, "only as subjects for laughter."

If, some five hundred years ago, an individual had submitted to the astronomical authorities a paper giving the results of his experiments with the pendulum, as a proof of the earth's daily rotation, those gentlemen would undoubtedly have stated that they did not agree with the conclusions arrived at from such experiments, and they might probably suggest that the author of the paper could not understand the perfection of the Epicycles of Ptolemy, which proved that the earth could not move.

Such proceedings in ancient times can be understood, but we have a far more difficult problem to deal with at the present day. We have now various learned societies, formed specially for the purpose of inquiring impartially on any problems connected with the special science for which the society was organized. The history of the past teaches us that erroneous theories were accepted as grand truths by all the scientific authorities of the whole world during more than five thousand years; and to these gentlemen it appeared that nothing more absurd and impossible than the daily rotation of the earth could be submitted to them.

We have, however, in the present day, a defect as regards the so-called scientific training of students, which ought at once to be remedied.

THE NAUTILUS CURVE.

In our schools and colleges, the science of geometry is taught, the earliest text-book usually being Euclid. From that book certain laws are made known relative to circles, ellipses, other curves, straight lines, and points. We there learn that the centre of a circle cannot vary its distance from the circumference; that a curve cannot increase or decrease its distance from a point at a uniform rate, and that the variation in the rate cannot be uniform.

According to certain popular theorists, all this geometry is wrong. The centre of a circle can, and does, vary its distance from the circumference, and yet always remains the centre of the same circle; for does not the pole of the heavens trace a circle round the pole of the ecliptic as a centre, and yet decreases its distance from this centre about 46" per century?

A curve can, and does, decrease its distance from a point at a uniformly increasing or decreasing rate; for was not the gold medal of the Royal Astronomical Society given for a paper on the supposed proper motion of the fixed stars, in which the assumed constant increase and decrease in rate was the base of the conclusions?

The laws of geometry teach that the extension of the arctic circle and tropics on a planet are dependent on the angle which the axis of daily rotation of that planet makes with the plane of its orbit; that if, from any cause, say a slight alteration in the position of the centre of gravity of the planet (such as the transfer of the waters of the oceans from one hemisphere to the other), the axis of the planet altered the angle it made with its orbit, then there would be a variation in the extent of the arctic circle and in the tropics.

The movements of the planet itself appear to the geometrician the important question for investigation, not the movement of the plane of its orbit.

But the movement which does or may take place in a planet cannot in any way affect the angle which the axis of this planet makes with its orbit, for has not M. La Place stated that, as the plane of the ecliptic cannot vary more than 1° 21', therefore no change greater than 1° 21' in the extent of the arctic circle ever did or ever can occur ?

To the mere reasoner acquainted with geometry, it appears that if the axis of Jupiter has "a conical movement" round a point only 10° from the pole of daily rotation, there would be, during one complete conical movement, a variation of 20° in the arctic circle, although the plane of the orbit of Jupiter did not vary one second. But this must be an error of geometry, because the French theorist has asserted that no change greater than 1° 21' can occur in the arctic circle, because the plane of the ecliptic cannot vary more than that amount.

Hence we are led to conclude that no matter how a planet moves, yet any such movement produces no effect on the extent of the arctic circle or tropics, unless the plane of that planet's orbit moves; whereas geometry teaches quite a different law.

To the mere reasoner, it appears by no means improbable that, because the obliquity of the ecliptic was the problem for inquiry, the great French theorist became a little mixed about the geometrical laws, and imagined that no change in this obliquity could occur except by a change in the ecliptic. He seemed to overlook the fact that no change whatever could occur in the obliquity, no matter how much the plane of the ecliptic varied, if the theory were correct that the pole of the heavens always traced a *circle* round the pole of the ecliptic *as a centre*.

The subject, however, which requires most serious attention is, whether students should continue to study Euclid and geometry.

THE NAUTILUS CURVE.   161

Many years are now devoted to these subjects, which are supposed to give knowledge of sound laws, immutable and unchangeable. If, however, when we come to such a science as astronomy, these laws are to be treated with contempt; if a paper containing a geometrical impossibility receives the gold medal of a learned society; if one of the asserted movements of the earth contains another geometrical impossibility; if the accepted assertion of a theorist interferes with another geometrical law;—then a knowledge of geometry must be a great detriment to a student who wishes to gain the favour of the present reigning scientific authorities.

If he wish to be accepted as a learned man, he ought to ignore geometry and its laws, and should learn by rote the assertions of admitted authorities, and bow to these in the same manner in which the astronomers during 1400 years bowed to the assertions of Ptolemy, and his feeble-minded copyists.

Any young man who may be desirous of gaining the approval of those persons, who, holding highly paid official scientific positions, have consequently considerable patronage, would really ruin his prospects if he were more convinced of the accuracy of geometry than he was of the theories of these authorities. It is, consequently, a serious question for the consideration of those gentlemen who regulate public education and examinations, whether geometry should not now be struck out, and the theories of authorities be substituted in its place.

On those persons who can prove the accuracy of the laws of geometry, the assertions and theories of those persons, which are opposed and contradicted by these laws, produce the same effect as when statements are read, "that it was utterly impossible for a steamship to cross the Atlantic;" "that if the earth rotated, all the water would

M

be flung off;" "that if the earth were spherical in form, the people underneath would fall off, as it was impossible for a man to walk head downwards on the ceiling," etc. The gentlemen who made these objections were all authorities in their day, and no doubt meant well; but, unfortunately, they did not know. They succeeded for a time in burking truth, but only for a time, and as it has been in the past, so it is only reasonable to assume it will be in the future.

We have, however, at the present time to make a selection: we must reject those theories, no matter from whom they emanate, which are contradicted by the laws of geometry, or we must reject geometry as a false and misleading science. The option is left to the reader.

So baneful, however, is the influence of mere authority, that there are a multitude of unreasoning individuals who would more readily believe in the infallibility of the theories of M. La Place, however much these might be contradicted by the laws of geometry, than they would accept as correct the accurately proved problems in Euclid.

CHAPTER XII.

THE ZENITH AND THE MERIDIAN.

THE belief which has prevailed among observers during the past two hundred years appears to be, that it is only necessary to know how much the pole of the heavens changes its position in the heavens during the year, and then, as the meridian must, from its very name, pass through the pole, the meridian zenith distance of a star will enable the declination or polar distance of this star to be immediately deduced therefrom.

This belief indicates that it is assumed that whatever change occurs in the direction and movement of the pole, must occur in each zenith and meridian. It also assumes that, no matter where a locality may be situated on earth, yet the zenith and meridian of this locality will be affected by "a conical movement of the earth's axis" (whatever such a vaguely defined movement may mean) exactly in the same manner as the zenith and meridian of any other locality is affected. These assumptions are not only without any foundation, but are erroneous, and they have been assumed mainly because it has been taken for granted that "a conical movement of the earth's axis" was a full and satisfactory explanation of that which really caused the pole of the heavens to change its position in the heavens.

When it is remembered that the earth is the actual

instrument with which observations are taken, it will be evident that it is essential that we know every detail of the movements that occur in this instrument. Let us refer to the following diagram (Fig. 52) as an example.

N Q S E represents the earth; N and S, the north and south poles; E Q, a portion of the equator; N A, N B, N C, etc., various meridians of terrestrial longitudes.

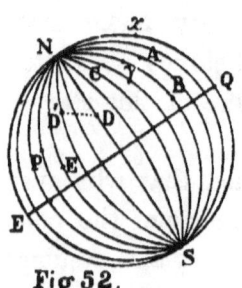

Fig 52.

"The earth's axis has a conical movement," is the present accepted theory.

If the axis does trace a cone, either the south point S or the north point N must remain fixed, whilst the pole which was not fixed described the base of the cone. Which pole remained fixed, and which pole described the base of the cone, has hitherto not been mentioned by theorists. It was, of course, thoroughly believed that the joint action of the sun and moon produced this conical movement of the axis, but perhaps a geometrician might imagine that, as a preliminary step, it would have been a more sound proceeding to define what the movement really was, than to invent a supposed cause for some vague movement. If the earth's axis trace a cone, as has been asserted, then one pole must remain fixed, whilst the other gyrates. This is sound geometry, and as the truth of this science rests on a firmer base than the assertions of theorists, it must hold a stronger position.

If neither pole remain fixed, whilst the two poles describe circles in the heavens, then the axis cannot trace a cone, but will trace two cones, provided that the apex of each is within the earth, or a portion of a frustrum of a cone, provided the apex is outside the earth.

From the manner in which this supposed movement has

been described by authorities, and the diagrams given to explain to their readers what the movement really is, it appears to have been assumed that the south pole remains fixed whilst the north pole gyrates. A peg-top or teetotum is referred to as an illustration of the movement of the earth, whereas there is no similitude between the two.

Taking, however, the theory as at present accepted by authorities, the following questions become of great and important interest.

Refer to the last diagram, and suppose that the earth's axis has performed 10° of its conical movement, so that the pole N has moved over 10° of its circular arc. Then mark on the sphere the position of the various zeniths A, B, C, D, etc., as affected by this change in direction of the earth's axis of 10°.

Will the zenith D be carried directly up the meridian towards N, or will it be carried in any other direction? Will the zenith B be carried directly towards N, or obliquely in some other direction? In fact, mark on the sphere the exact positions of these zeniths as affected by "a conical movement of the earth's axis."

It is almost unnecessary to state that the solution of this problem is beyond the knowledge of any professional astronomer. If he mark the position which he imagines any zenith will occupy, he at once gives a definite movement to the earth. If he assert that the zenith D moves up the meridian D N, he must claim that the south pole remains fixed during the conical movement. If he claim that the zenith D is carried to the left, he must claim that some other movement of the earth occurs in connection with the conical movement; and what is this other movement.?

Are we to leave so important a problem as the movement of the earth—the actual instrument with which we observe—vague and undefined, and then, by employing at great expense numbers of observers and computers, continue to muddle on for a few years in advance, by aid of perpetual observations, in order to frame an astronomical almanac for two or three years in the future? Does not the very fact of the necessity of these perpetual observations prove that the results for the future cannot be calculated by the simple laws of geometry?

The effects on each zenith and meridian of the second rotation of the earth will now be defined, and the reader will then perceive how very simple these changes become, but how hopelessly complicated they would appear to those persons who were unacquainted with this fact.

Take a point, x, on the sphere (last diagram, Fig. 52) 29° 25′ 47″ from the pole N, the point x being the pole of the second axis of rotation.

The pole N of daily rotation is carried round x as a centre, and each zenith, A, B, C, D, etc., is carried round x as a centre. Whilst moving round x as a centre, the zenith B is carried nearly parallel to the equator, E Q, but the zenith y on the same meridian is carried nearly at right angles to the equator, this zenith y being also carried round x as a centre.

How, then, is the readjustment of this change to be effected? It is accomplished by the *daily* rotation of the earth—a movement which affects the apparent right ascension of stars, according to the present system of measuring right ascensions.

This change in the meridian and zeniths due to the second rotation, and hitherto unknown to theorists, is one of the causes which produces effects hitherto erroneously attributed to "a proper motion" in the stars themselves.

In each case a point on the earth's surface is carried over an arc round x as a centre; the pole N is carried over

THE ZENITH AND THE MERIDIAN. 167

an arc of about 20·09" annually, but a point on the earth's surface 90°· from x will be carried annually over an arc of 40·9".

The direction of the arc over which the pole N is carried is approximately towards the first point of Aries, but the direction in which a point D is carried is towards D', and the arc D D' is greater than the arc over which the pole N is carried, because the distance D x is greater than the distance N x.

A point on the meridian N A, so situated that an arc from x to this point is at right angles to the meridian N A, will be carried directly up this meridian, consequently towards N, the position of the pole, whilst a point on the equator of this same meridian will be carried at right angles to the arc joining x to this point.

As the centre of gravity of the earth remains fixed *as regards* both the daily and second rotation, the variations in position of different localities on earth, as produced by the second rotation, causes a marked difference in the position, not only of the zeniths, but also in the horizon of various localities. How varied are these differences may probably be better understood by means of a careful examination of the following diagram (Fig. 53).

P represents the north pole of the earth at a date in the past; T S R Q O V, the circle traced by the zenith of a locality, say, in 51° north latitude during one daily rotation.

The circle F E B A G represents the circle traced during the same daily rotation by the zenith of a locality, say, in 68° north latitude.

x represents the pole of the second axis of rotation, round which the pole P and the various zeniths describe circles during one second rotation.

P T, P S, P R, etc., represent various meridians.

Any number of years after the pole of daily rotation was at P, it is carried to P' round x as a centre, the direction of P P' being nearly towards the first point of Aries—that is, nearly towards 360° of right ascension.

No movement of the earth can alter the angular distance between a given locality and the poles, as long as the axis remains fixed *in the earth.*

With P' as the new position of the pole of daily rotation, we can trace two circles, represented by the dotted

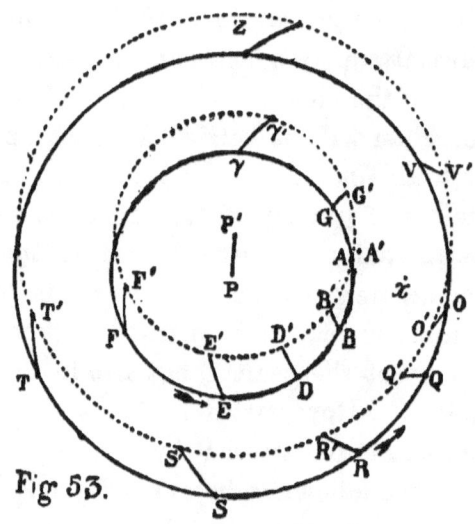

Fig. 53.

lines, to mark the circles traced during a daily rotation by the zeniths of the two localities before mentioned, the angular distance of P' from the dotted circles being equal to the angular distance of P from the continued circles.

We can now trace out and calculate the exact effects on various zeniths produced by the second rotation of the earth.

The pole P is carried to P' round x as a centre; the arc P x = P' x = 29° 25' 47".

It must be remembered that if a person were located

THE ZENITH AND THE MERIDIAN. 169

exactly at P, the north pole, *his zenith* would be transferred from P to P' on the sphere of the heavens.

The zenith which was at T will be transferred to T', the arc T T' being traced round x as a centre. The zenith S is transferred to S', R to R', Q to Q', O to O', V to V', etc.

The zenith F of the higher latitude is transferred to F', E to E', D to D', B to B', A to A', and G to G', each of these small arcs having the point x as a common centre.

The meridian which, when the pole was at P, traced the arc P F T, is transferred by the second rotation to P' F' T'. The meridian P E S is transferred to P' E' S'. The meridian P D R is transferred to P' D' R', thus intersecting the position of the former meridian.

All the meridians from P S round by P O to P Z will intersect each other in consequence of the second rotation; that is, the meridian of say fifteen hours for 1887 will intersect, near the pole of daily rotation, the meridian of fifteen hours for 1886.

In the direction from 0 hours to six hours, and from six hours to twelve hours, the meridian of one year will not intersect the meridian of the year previous.

Between about sixteen hours right ascension and twenty hours right ascension the most marked changes will occur, in consequence of the second rotation.

In order to understand these changes, the reader must bear in mind that the second rotation *is in opposition* to the daily rotation. For example, if a mere conical movement of the axis occurred, then whilst the pole P was carried to P', the zenith S would be carried up the meridian towards N. When, however, the second rotation is comprehended, it will be seen that, whilst the pole is carried from P to P', the zenith S is carried to S', and it has to regain the former position by means of the *daily* rotation.

The zenith D, however, is carried by the second rotation

to D', over an arc which coincides with the meridian, R D P.

In the immediate vicinity of x (the pole of second rotation), the changes in the zenith, both as regards direction and amount, will be very variable. All points on the earth's surface between x and P will be carried by the second rotation in the same direction as they are by the daily rotation, whereas all points on the earth's surface between x and the equator will be carried by the second rotation in a direction opposite to the daily rotation.

There will also be remarkable changes in or near meridians of six hours right ascension.

By the daily rotation, an arc of the equator or equinoctial joining two meridians will be greater as effected by the daily rotation than will be an arc on these two meridians joining two zeniths at a distance from this equator.

A zenith, say 29° 25′ 47″ north of the equator of slow rotation, would be carried annually by the second rotation over an arc, in opposition to the daily rotation, exactly equal to the arc over which a zenith 29° 25′ 47″ south of the equator of slow rotation would be carried. But that part of the meridian of six hours right ascension which is 29° 25′ 47″ south of the equator of second rotation, is on the equator of *daily* rotation.

Hence, a zenith of a locality in latitude 59° 51′ 34″ north would be carried annually by the second rotation over exactly the same arc as would that portion of the meridian of six hours right ascension which was on the equinoctial.

Between the equator at six hours right ascension and the zenith of a locality 29° 25′ 47″ north of the equator of second rotation, the meridian would be carried over an arc by the second rotation greater in amount than was this zenith on the equator of daily rotation.

Hence, when we come to stars in this part of the

heavens, we encounter a very marked condition of confusion when their positions have to be determined by perpetual observation, and their right ascensions by their meridian transits. Consequently, theorists have attributed to these stars endless variations of proper motions, which so-called proper motions are in the zeniths and meridians, not in the stars.

A very fair example of this law is afforded by the star a Aurigæ (Capella).

The constants for this star are as follows (Fig. 54): P C = 29° 25' 47". From C, the pole of second rotation, to P, the pole of daily rotation, C a = 73° 2' 39·5", a being the star. Angle P C a, January 1, 1887 = 9° 21' 39·8". Annual decrease (apparent) in the angle P C a = 37·01".

Results:—

Date.	Mean polar distance by observation.	By calculation.
January 1, 1887		
,, 1873	44° 8' 3·04"	44° 8' 3"
,, 1850	44° 9' 39·4"	44° 9' 39·5"
,, 1843	44° 10' 7·42"	44° 10' 8·6"
,, 1819	44° 11' 53"	44° 11' 52·2"
,, 1780	44° 14' 43"	44° 14' 44·8"

The varied effects which are produced on the meridian zenith distance of a star as seen below the pole and above the pole, and also its time of transit across the meridian, afford some very interesting problems.

For example, suppose a star to be near the meridian F P G, and near the pole P of daily rotation. Each year the zenith F is carried over a much longer arc by the second rotation, in opposition to the daily rotation, than is the zenith G. Consequently, when right ascensions are counted from an imaginary standard of time, such as the successive transits of the first point of Aries, we obtain some singular confusion in this item.

The principal cause, however, of the confusion which now exists, and which necessitates the continuous observations now supposed to be essential, is due to the following fact.

A meridian, although an arc of a great circle, appears to an observer as a *straight* line passing through the pole and the zenith, and cutting the equinoctial and horizon at right angles. The transit instrument used for measuring zenith distances and meridian transits moves in this imaginary straight line.

If the movement of the earth were merely a change in the direction of the axis, without a second rotation, there

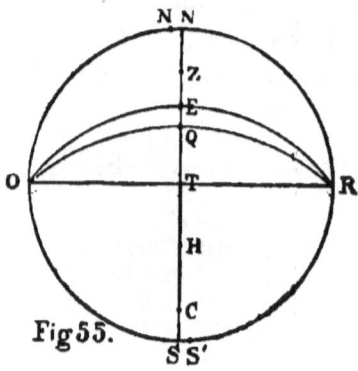

Fig 55.

would be less confusion than at present exists; but when the second rotation occurs, we have the changes in every meridian.

As an example of one form of these changes, a meridian of six hours right ascension will be referred to, and reference made to a locality on earth in 51° north latitude.

The circle N O S R (Fig. 55) represents the earth; N the north, S the south, pole; O T R, the equator; Z, a locality in 51° north latitude; H, the southern point on the horizon 90° from Z, as seen on the sphere of the heavens.

O Q R represents the trace of the ecliptic; O E R, the trace of the equator of slow rotation.

At the end of a year, if no change in direction of the earth's axis occurred, the meridian N Z T H S would, after say 366 complete sidereal rotations of the earth, coincide with its former position, viz. N Z T H S.

But during one year a change does occur in the direction of the axis; the pole N is carried to N', and the pole S is carried to S', the arcs N N' and S S' being each equal to about 20·09" for each year.

The points N and S, although spoken of as the poles of the earth, are merely points on the earth's surface unaffected by the *daily* rotation. Both these points have a zenith, and this zenith changes its position in the heavens about 20·09" annually.

But the north and south poles are not the only points on the earth's surface that have zeniths which are affected by *some* movement of the earth, which movement is independent of the daily rotation. In what manner is the zenith of Z affected? In what manner are the zeniths of E, Q, T, and H affected by that movement of the earth which causes the pole N to be transferred to N', and S to S'?

Strange as it may appear to any reasoner, this is a question in astronomy which has never been dealt with. We have any number of visionary imaginings presented to us by theorists as regards the supposed cause of *some* movement of the earth, but what this movement really is has never hitherto been defined.

Although, from the most simple observations, it is known that the pole N is carried to N' over an arc of about 20·09" annually, it has never even been hinted as to the amount or direction in which the zenith of latitude 51° north has been carried, or the amount and direction in which the zenith of any other locality is annually carried.

Why is there this omission from a science claimed to be so exact as is astronomy? The answer is very simple.

Theorists do not know what the movement of the earth really is which causes the zeniths of the poles to vary their position in the heavens about 20·09″ annually. If any theorist marked on the sphere the points to which the zenith Z and the zeniths E, Q, T, and H would be carried by the same movement which causes the zenith N to be carried to N′, and S to S′, he would at once commit himself to an exact movement of the earth, and as this exact movement has been hitherto unknown, it is more prudent to keep to such a vague assertion as "a conical movement of the earth's axis."

This detail movement of the earth, unimportant as it may appear to the average reader, represents a very large sum of money when considered from a mere business point of view. During about a hundred and fifty years, observations to determine the future position of stars have been made with the transit instrument, in the plane of the meridian; a large staff of observers and computers have been employed for this work only. Several observatories in the United Kingdom carry on this same work.

That such mere routine work must be totally unnecessary if, as was claimed, the true movements of the earth were really known, is an idea which never seems to have dawned on the mind of any reasoner. The amount of money paid at various official observatories, for persons engaged on this routine work, is understated when it is put down at £10,000 per annum. One hundred and fifty years of such work represents a sum of one and a half million pounds sterling, spent for making observations which must be unnecessary if the true movements of the earth were really known.

Hundreds of pages of the journal, and of the proceedings of scientific societies, have been filled with lists of the supposed proper motion of the fixed stars; the gold medals of scientific societies have been given for these papers, amidst the cheers of the fellows; and yet, strange to say, in no single instance have any one of these distinguished gentlemen thought it necessary ever to examine or define how the zenith and meridian of their place of observation was affected by that same movement, which caused the zeniths of the two poles to trace an arc of $20\cdot09''$ annually in a known direction. There are a number of individuals even in the present day who assert that they know how long the sun will last, how much fuel it consumes, what is the constitution of each star in the heavens, etc. It seems almost an insult to ask such gigantic intellects to descend from their thrones, and to define the actual movement of the earth produced by "a conical movement of the axis." So important, however, is this movement, so impossible is it to arrive at any sound knowledge by means of repeated observations unless the true movements of the earth are known, that, trifling as the request may appear, yet it may with justice be asked that some learned gentlemen mark on the sphere given in the last diagram, the amount and direction in which the zeniths of Z, E, T, and H are moved by that same motion of the earth which causes the pole N to be carried to N', and S to S'.

In the history of astronomy, we have a precedent why such trifling matters as the shape and movements of the earth are beneath the notice of grand intellects. The ancient astrologers, for example, claimed to have discovered the lucky or baneful influence of every star and planet on each human being, according to the time of his birth. It would scarcely be expected that these ancient gentlemen could condescend to turn their thoughts on so small a thing as the earth; consequently, though they claimed to know all about the distant stars and planets, they imagined that the earth was a flat surface, and that it never moved.

There may be, however, individuals in existence who can turn their attention to small matters, and for the information of these, the detail movements of the zenith and meridian will now be defined.

In the following diagram (Fig. 56) N O S R represents the earth; N the north, S the south, pole; O T R, the equator; O Q R, the trace of the ecliptic; O E R, the trace of the equator of second rotation; N Z Q T H S, a meridian of six hours right ascension; Z, a locality in 51° north latitude; E, Q, and T, the points where the meridian intersects

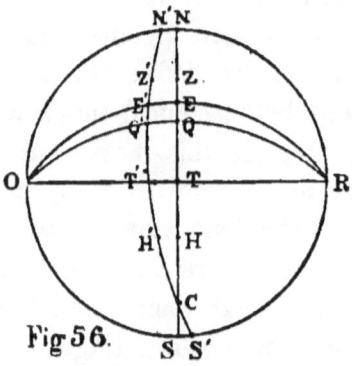

Fig 56.

the equator of second rotation, the ecliptic, and the equator of daily rotation.

H, the south point of the horizon projected on the sphere of the heavens 90° from Z.

The north pole of the second rotation impinges on the earth 29° 25′ 47″ from N, on the opposite side of the sphere. The south pole of the second axis of rotation impinges on the earth at C, 29° 25′ 47″ from the south pole S.

The second rotation occurs at the rate of 40·9″ annually, consequently the point E during one year is transferred to E,′ the arc E E′ being 40·9″.

The pole N, distant 90° − 29° 25′ 47″ from the equator of second rotation, is moved by the second rotation over an arc of 40·9″ × cosine of 60° 34′ 13″ = 20·09″ from N to N′.

The zenith of Z, a locality in 51° north latitude, and therefore 51° − 29° 25′ 47″ = 21° 34′ 13″ from the equator of slow rotation, is moved by the slow rotation over an arc of 40·9″ × cosine of 21° 34′ 13″ = 38″, from Z to Z′.

The point T on the equator, 29° 25′ 47″ from E, is moved by the second rotation over an arc of 40·9″ × by the cosine of 29° 25′ 47″ = 35·62″ = T T′.

The point H on the horizon for this latitude is distant from E, the equator of slow rotation, E T + T H = 29° 25′ 47″ + 39° = 68° 25′ 47″. The arc over which H is carried by the slow rotation is, therefore, 40·9″ × cosine of 68° 25′ 47″ = 15·03″ only.

The continuation of the meridian will pass through C, the pole of second rotation, and through S′, the pole of daily rotation.

We have, then, the following changes in the zeniths and meridian for six hours right ascension, and 366 daily rotations of the earth, as produced by the second rotation only:—

$$N N' = 20·09''$$
$$Z Z' = 38''$$
$$T T' = 35·62''$$
$$H H' = 15·03''$$
$$\text{and } E E' = 40·9''$$

Hence, the zenith for six hours right ascension, and for latitude 51° north, is carried by the second rotation annually, in opposition to the daily rotation, over an arc of 2·38″ greater than is the equator of daily rotation.

The meridian of one year also forms, with the meridian of a previous year, a curve, the most distant points of which are in latitudes 29° 25′ 47″ north.

The transit instrument which swept down the meridian N Z E T H C will, after 366 daily rotations of the earth, sweep down the meridian N′ Z′ E′ T′ H′ and C′, which again appears as a straight line.

For this meridian to be carried by the *daily* rotation so that the point T' comes again to T, the zenith Z' will by this *daily* rotation have failed to reach Z, because the equator is by the *daily* rotation carried over a greater arc than is the zenith of a locality in 51° north latitude; whereas by the second rotation the zenith of 51° is, for six hours right ascension, carried over a greater arc than is the equator of daily rotation for that meridian.

Hence, as the right ascension of stars are determined by the time at which these stars transit the meridian, there appears by the present system a discordance or difference in the increase of right ascension in two stars, which difference is due, not to any proper motion in the stars themselves, but to the effect produced on the zenith and meridian by the second rotation.

On meridians at or near eighteen hours right ascension, the effects of the second rotation is even more varied and remarkable than it is near six hours right ascension. The zeniths of some localities are carried almost directly towards the north pole, whilst the zenith of other localities on the same meridian are carried nearly at right angles to an arc joining these zeniths with the north pole.

These varied movements of the zeniths and meridians cause a confusion in right ascensions and zenith distances which the mere observer is quite incompetent to unravel. The staff of our official observatories may be increased by scores of observers and computers, and with no other result than that which followed the adding of additional epicycles to the system of Ptolemy, viz. making confusion more complicated. The solution of the mystery is very simple, and can be easily arrived at by those persons who can move out of the erroneous grooves in which they have during so many years been content to travel.

In the following diagram (Fig. 57), the changes produced by the second rotation on various zeniths for a meridian of eighteen hours right ascension will be described.

The circle N Q S R represents the earth; N the north, S the south, pole; Q T R, the equator of *daily* rotation; Q E R, the equator of second rotation.

N C Z T H S, a meridian of eighteen hours right ascension.

C, the pole of second rotation, 29° 25′ 47″ from N.

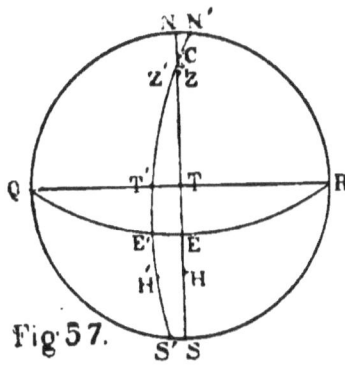

Fig 57.

Z, a locality in 51° north latitude.
T, a locality on the equator of daily rotation.
E, a locality on the equator of second rotation.
H, a point 90° from Z.

The most visible effect of the second rotation is to cause the pole N to be carried to N′, and S to S′, each of these arcs being 20·09″, and representing one year of the second rotation.

The effects on other zeniths will be as follows: Z will be carried to Z′ round C as a centre. The length of the arc Z Z′ can be found as follows: E Z = E T + T Z = 29° 25′ 47″ + 51° = 80° 25′ 47″. The effect of the second rotation is greatest at the equator of second rotation, where

180 UNTRODDEN GROUND IN ASTRONOMY AND GEOLOGY.

it is 40·9″ annually; the arc Z Z′ is therefore 40·9″ × cosine of 80° 25′ 47″ = 6·79″.

The arc T T′ of the equator of daily rotation is equal to 40·9″ × cosine of E T = 35·624″.

E E′ = 40·9″.

H H′ = 40·9″ × cosine of E H = 40·32″.

Hence the meridian of eighteen hours right ascension is affected by the second rotation only—by the same movement of the earth, it must be borne in mind, which causes the pole N to be carried to N′, and S to S′, as follows:—

C, the pole of second rotation remains fixed.

$$Z Z' = 6·79''$$
$$T T' = 35·62''$$
$$E E' = 40·9''$$
$$H H' = 40·32''$$
$$S S' = 20·09''$$

The meridian will, therefore, be displaced by the second rotation in the manner shown by the great circle passing from N′ to C, Z′, T′, E′, H′, to S′.

The arc of the equator intercepted between these two meridians will be T T′ = 35·62″, whereas the arc intercepted between the two zeniths Z Z′ = 6·79″ only. Consequently, when the *daily* rotation brings the point T′ to T, the zenith Z′ will be carried past its former position, Z, by the daily rotation; consequently, it will appear as though this zenith had been displaced to a greater amount by the second rotation than it really had been affected.

Such a star as γ Draconis, with a right ascension in 1887 of 17h. 53m. 58·88s., and a north declination at the same date of 51° 30′ 8·4″, consequently near the zenith Z, serves well to illustrate this effect.

The constants for this star are as follows (Fig. 58):—

$$P C = 29° 25' 47''.$$
$$C \gamma = 9° 6' 22·5''$$
Angle P C γ, January 1, 1887 = 174° 4′ 19·7″.
Annual rate of variation in angle C, 45″.

Fig. 58.

The following are the results obtained by calculation compared with those recorded as having been obtained by observation at various dates for the mean polar distance :—

Date.	Calculation.	Recorded.
January 1, 1887	38° 29′ 51·7″	38° 29′ 51·6″
,, 1873	38° 29′ 43·5″	38° 29′ 43·9″
,, 1850	38° 29′ 29·4″	38° 29′ 29·4″
,, 1780	38° 28′ 43″	38° 28′ 46″

For a period of 107 years, viz. from 1887 back to 1780, the polar distance of this star can be calculated, and a difference of only 3″ exists between the calculation and the recorded observation at the earlier date. Considering the vague refraction used at the date 1780, and also the imperfection of the instruments then used, it is for the reader to judge which is the more likely to be correct, calculation, or observations made 107 years in the past; the difference, however, is so very slight as to serve as a fair example of the calculations that can be made when the second rotation of the earth is comprehended.

Another star near the meridian of eighteen hours right ascension serves as an example of the changes in the meridian in this part of the heavens as effected by the second rotation. This star is Vega α Lyræ.

The constants for this star are as follows (Fig. 59) :—

P C = 29° 25′ 47″.
C α = 22° 29′ 51·2″.
Angle P C α, January 1, 1887 = 162° 55′ 6·8″.
Annual variation in this angle, 44·4″.

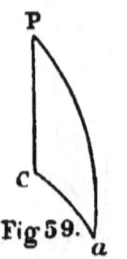

Fig 59.

In order to refresh the reader's memory as regards the method of calculating the mean polar distance for more than one hundred years, a calculation is given below for calculating the mean polar distance of α Lyræ for January 1, 1755.

The interval between 1755 and 1887 is 132 years, which, multiplied by 44·4″, amounts to 1° 37′ 40·8″.

This angle added to 162° 55′ 6·8″ amounts to 164° 32′ 47·6″, which will be the angle at C for 1755.

With the two sides P C and C α, and the included angle at C, we can calculate P α, the polar distance of α Lyræ for 1755, by the usual formula, as follows:—

The supplement of 164° 32′ 47·6″ is 15° 27′ 12·4″.

Log. cosine, 15° 27′ 12·4″ = 9·9840083
Log. tangent, 22° 29′ 51·2″ = 9·6171719

9·6011802 = log. tan. first arc, 21° 45′ 41·1″.

29° 25′ 47″
21° 45′ 41·1″
─────────
51° 11′ 28·1″ = second arc.

Log. cosine, 22° 29′ 51·2″ = 9·9656230
Log. cosine, 51° 11′ 28·1″ = 9·7970765
─────────
19·7626995
− Log. cosine, 21° 45′ 41·1″ = 9·9678921
─────────
9·7948074 = log. cos. P α = 51° 25′ 51·2″.

In Bradley's catalogue for 1755, the polar distance of α Lyræ is given, as observed by him, as 51° 25′ 49″, showing a difference of 2·2" only between Bradley's observation, made 132 years in the past, and the result arrived at by calculation. Here again we must consider the class of instrument used by Bradley, and the uncertainty of the refraction he used to correct his zenith distances, and it is again left to the reader's judgment to consider which is the more dependable, calculation or observation.

By the same means as shown above, the calculated polar distance of a Lyræ, compared with that recorded as found by observation, is as follows :—

Date.	Calculation.	Recorded.
January 1, 1887 ..	.. 51° 19' 16" ..	.. 51° 19' 16"
„ 1850 ..	.. 51° 21' 10·8" ..	.. 51° 21' 10·0"

When it is borne in mind that hitherto no method has been known to astronomers by which they could *calculate* the polar distance of a star for even five years, but were obliged to employ a mere rule-of-thumb method of adding or subtracting an annual rate, found merely by perpetual observation, and then employing this for two or three years only, it may possibly be admitted that the real calculation given above is a considerable advance, and possesses the great advantage of simplicity, and renders unnecessary, perpetual observations.

When a point on the earth's surface is so situated that the arc from the zenith of this locality to the pole of the second axis of rotation forms a right angle with the arc drawn from the pole of daily rotation to the pole of second rotation, then this zenith will be carried by the second rotation directly towards, or away from, the pole of daily rotation according as the meridian of right ascension is greater or less than eighteen hours.

A very slight difference in the polar distance of a star will cause a very great difference in the zenith of that locality in which the star is situated, the zenith being

184 UNTRODDEN GROUND IN ASTRONOMY AND GEOLOGY.

carried in one instance directly towards that position in the heavens occupied by the pole of daily rotation; in the other case, the zenith is carried nearly at right angles to the arc joining that zenith with the pole of daily rotation.

As these zeniths alter their positions considerably from year to year in consequence of the second rotation, a constant rate cannot be used according to the present system of determining polar distances from zenith distances.

As an example of these apparent mysteries, two stars will be given, viz. ζ Ursæ Minoris and η^2 Draconis.

The constants for ζ Ursæ Minoris are (Fig. 60)—

P C = 29° 25′ 47″.
C ζ = 20° 25′ 59·2″.
Angle P C ζ, January 1, 1887 = 18° 41′ 3″.
Annual variation in angle P C ζ = 41″.

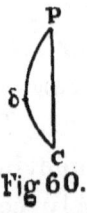

Fig. 60.

The results by calculation compared with recorded observation are, for polar distance—

Date.	Calculation.	Observation.
January 1, 1887	11° 51′ 31·3″	11° 51′ 31·3′
„ 1850	11° 44′ 48·1″	11° 44′ 48″

For the star η^2 Draconis, the constants are as follows:—

P C = 29° 25′ 47″.
C η^2 = 11° 44′ 57″.
Angle P C η^2, January 1, 1887 = 73° 32′ 35·6 .
Annual variation in angle, 40·72″.

Fig. 61.

THE ZENITH AND THE MERIDIAN. 185

The following are the results obtained by calculation, compared with those recorded for the polar distance of this star:—

Date.	Calculation.	Recorded.
January 1, 1887	28° 13' 48·4"	28° 13' 48·4"
„ 1873	28° 11' 52·7"	28° 11' 52·67"
„ 1850	28° 8' 42·4"	28° 8' 42·4"

The difference in the effect of the rate of these two stars, as produced by the influence on the two meridians by the second rotation, amounts during thirty-seven years to about 0·28" annually. The amount 0·28" annually, measured on the equator of daily rotation and converted into time, is less than one-tenth of a second of time per year—a rather small amount to be checked by any chronometer, but yet perceptible when an interval of several years is dealt with.

The changes in the various zeniths and meridians in this part of the heavens, as produced by the second rotation, are so numerous and so varied, that calculations have to be made for each zenith or star as affected by this movement. These apparent complications are not due, however, to the assertion that the whole solar system is rushing towards the constellation Herculis, at the rate of upwards of one hundred and fifty-four million miles per year; nor are they due to the assertion that the stars themselves are all rushing about. The effect of the second rotation of the earth on the meridians and zeniths in this part of the heavens, in a manner with which theorists have not been aware, is the true cause.

When we come to meridians of twelve hours or 0 hours of right ascension, we do not meet with those differences in the effects of the second rotation which are manifested from about fourteen to twenty-two hours of right ascension; the reason for this is almost manifest.

Meridians at or near to 0 hours and twelve hours of right ascension are displaced by the second rotation nearly in opposition to the daily rotation, and in such a manner that when the *daily* rotation carries these meridians over the small arc through which they were carried by the second rotation, the meridian of one year almost coincides with the position it previously occupied.

The following diagram (Fig 62) will explain this fact:—

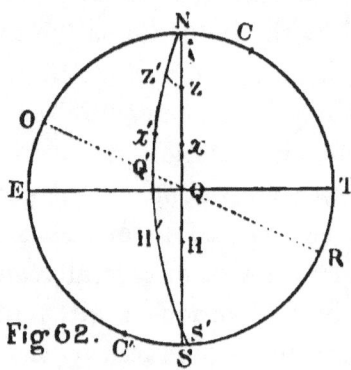

N E S T represents the earth; N the north, S the south, pole; E Q T, the equator of daily rotation.

C represents the position of the pole of the second axis of rotation; C', the opposite pole of second rotation; O Q R, the equator of second rotation; O E = N C = 29° 25' 47".

N Z Q H S represents the meridian of twelve hours right ascension; Z, a locality in 51° north latitude; N Z therefore = 39°; H, a point 90° from Z; x, a point on the same meridian, in latitude 20° north.

The effect of the second rotation on this meridian during one year is as follows:—

The pole N is carried over an arc of 20·09", round C as a centre. Consequently, the pole N is carried *down*, as we may term it, and in the direction of 0 hours right ascension, as shown by the arrow near N.

The pole S will be carried to S', round C' as a centre; the arc S S' = 20·09".

The point Q on the equator of second rotation will be carried to Q'; the arc Q Q' = 40·9".

The zenith Z will be carried to Z', the arc Z Z' being a portion of a small circle having C for its centre. The value of the arc Z Z' can be calculated as follows:—

N Z = 39°. N C = 29° 25' 47"; angle at N a right angle. The hypotenuse C Z is therefore, in round numbers, 47° 23'.

The point Z is therefore, 47° 23' from C, and 42° 37' from the equator of slow rotation.

The extent of the arc Z Z' will therefore be 40·9" × cosine 42° 37' = 27·69".

The value of the arcs $x\ x'$ and H H' can be calculated in the same manner, and the displacement of this meridian by the second rotation can be traced out in all its details, as shown by the great circle drawn from S to H', Q', x', and Z'.

It will be evident that the displacement of this meridian by the *second* rotation is almost identical with the displacement caused by the *daily* rotation. Consequently, the daily rotation readjusts this meridian very nearly, although the zeniths H, Q, x, and Z will have been displaced 20·09" up this meridian during one year.

The great displacements in the various zeniths and meridians which occur near eighteen hours right ascension, in consequence of the second rotation, do not occur near twelve hours or 0 hours right ascension. Hence the apparent rates of stars near these meridians vary very slightly, as will be seen by the following examples.

The stars λ Draconis, α Draconis, α Ursæ Majoris, γ Ursæ Majoris, η Ursæ Majoris, δ Leonis, β Leonis, α Canum Venaticor, ε Virginis, Spica Virginis, and β Corvi, vary in their annual rates as affected by the second rotation, between 40·76" per year and 41·1".

Take, for example, the star β Leonis, and calculate the polar distance of this star for any date, and it will be found that calculation gives a rate which is in accordance with the laws of geometry, whilst recorded observations, in many instances, give a sudden change in the rate, then a return to the before-mentioned rate, and so on.

The following are the constants for β Leonis (Fig. 63):—

P C = 29° 25' 47'.
C β = 78° 49' 14·6".
Angle P C β, January 1, 1887 = 78° 49' 43·6".
Annual variation in angle = 41·0".

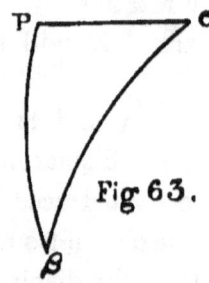

Fig 63.

RESULTS OF POLAR DISTANCE, P β.

Date.	Calculation.	Recorded.
January 1, 1887	74° 47' 46·8"	74° 47' 46·8"
,, 1850	74° 35' 23·5"	74° 35' 22·6"
,, 1755	74° 3' 36·6"	74° 3' 35·6"

The star γ Ursæ Majoris is another example. This star is north of the zenith of 51° north latitude, and the second rotation affects the meridian slightly less than it does the meridian on which β Leonis is situated.

The constants for γ Ursæ Majoris are as follows (Fig. 64):—

P C = 29° 25' 47".
C γ = 46° 11' 0·4".
Angle P C γ, January 1, 1887 = 53° 49' 3".
Annual variation in angle P C γ = 40·8".

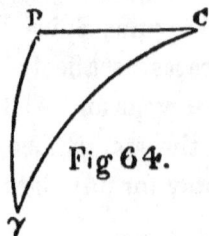

Fig 64.

THE ZENITH AND THE MERIDIAN. 183

The following is a comparison between the polar distance found by calculation and recorded at various dates:—

Date.	Calculation.	Recorded.
January 1, 1887	35° 40' 37·2"	35° 40' 37·2"
,, 1873	35° 35' 57·2"	35° 35' 57·4"
,, 1850	35° 28' 16·8"	35° 28' 16·8"
,, 1830	35° 21' 35·1"	35° 21' 36·5"
,, 1755	34° 56' 36·7"	34° 56' 35"

In the various nautical almanacs professing to give the declination of the stars, this declination is given to the one-hundredth of a second, whilst the annual rate of change in this declination is given to the one-thousandth of a second. Any practical observer who is acquainted with the uncertainty of refraction, knows that this asserted minute accuracy is theoretical only, not really a fact. Whilst, however, this accuracy is claimed, we frequently come across some amusing descrepancies, one of which may be mentioned in connection with the star γ Ursæ Majoris.

In the Nautical Almanac for 1873, the mean declination of this star for January 1 is given as 54° 24' 2·51"; in the Nautical Almanac for January 1, 1887, 54° 19' 22·80"; giving a difference for fourteen years of 4' 39·71". That is a decrease in the declination of 279·71" in fourteen years, which is at the rate of 19·979" per year. If arithmetic is correct, this rate of 19·979" per year causes the declination given for January 1, 1873, to become 54° 19' 22·80" on January 1, 1887.

But in the Nautical Almanac for 1873 it is stated that the annual variation in declination for this star is 20·024", and in the Nautical Almanac for 1887 it is stated that the annual variation is 20·026".

Which, then, are we to accept as correct, the recorded declinations as given in the Nautical Almanac and the rules of arithmetic, or the theories, formulæ, and rate, also given as correct? It is claimed that the annual rate for this star is known to the thousandth of a second. How is it that the star refuses to conform to this rate? The explanation of the theorist is, of course, very simple: the theory cannot move, so the star must do so, and γ Ursæ Majoris must have a proper motion.

As an example of the far greater accuracy to be obtained by calculation than can be reached by mere observation, it must be borne in mind that when the second rotation of the earth is understood, calculation can be made in order to find the polar distance or declination of a star for any date quite independent of the annual rate of change in these items found by observation.

The geometrician may, as a matter of mere amusement, find what the mean rate was between any two dates, by dividing the difference between the polar distances found by calculation by the number of years between the two dates.

For example, calculate the polar distance of this star for January 1, 1887, and January 1, 1850, and we obtain as follows:—

January 1, 1887 = 35° 40' 37·2"
„ 1850 = 35° 28' 16·8"
Difference for thirty-seven years = 12' 20·4"
Being at the rate of 20·01" annually.

Again, calculate the polar distance of this star from January, 1 1830, and compare this calculation with that obtained for 1850, and we have as follows:—

January 1, 1850 = 35° 28' 16·8"
„ 1830 = 35° 21' 36·5"
Difference for 20 years = 6' 40·3"
Being at the rate of 20·01" per year.

The annual rate given for this star in star-catalogues for 1850 was 20·04", whilst in 1887 it was 20·026".

Hundreds of similar examples could be given, proving

that the minute accuracy claimed by theorists, such as giving the rates of stars to one-thousandth of a second, is theoretical only. The recorded observations alone prove this assertion to have no foundation in fact.

A multitude of examples could also be given, proving that observations cannot compete in accuracy with calculations—that is, unless we are prepared to admit that arithmetic and geometry are not so much to be depended on as the theories of certain authorities.

It will now probably be evident to the geometrician and reasoner, that the manner in which each meridian and each zenith is affected by the second rotation, is a problem of the greatest importance when observations are made by the transit instrument on the present system.

In the whole history of modern astronomy, whether we examine the various books that profess to deal with the science of astronomy, or with the details of the calculations, not one word is written as regards the changes that occur in connection with various zeniths and meridians, and produced by "a conical movement of the earth's axis." In the numerous papers published in the monthly notices of the Royal Astronomical Society, the subject is never referred to. The most vague and baseless theories, if popular, are given many pages, but so important an item as how the zenith and meridian are affected, seems, if dealt with in a paper, to be considered only suitable for "a pigeon-hole."

There are several problems connected with the second rotation which require to be worked out, more in detail, than one person alone could work out in a lifetime. But the most mysterious problem of all is to explain why so important a problem as the changes in the zeniths and meridians is considered beneath the notice of men claiming to be scientific.

At present the daily rotation and the second rotation are mixed up together, and confusion consequently ensues. To clear up this confusion, it is necessary to examine the effects of the second rotation by itself, and then to note in what manner the daily rotation is interfered with by the movements given to the earth by the second rotation. Hitherto the only notice taken of the second rotation is that the zenith of the two poles of daily rotation alter their positions in the heavens about 20·09" annually; all else has been ignored.

CHAPTER XIII.

THE MEASUREMENT OF TIME, AND RIGHT ASCENSION.

IT is a most unusual proceeding to find an authority admitting that there is anything connected with his special subject which is not known. The very opposite proceeding is usually practised, by claiming that everything *is* known, and that the special science of which the individual may, for the time being, be a popular authority, has been exhausted.

It speaks well, therefore, for the candour of the late Sir John Herschel that, in his "Outlines of Astronomy," in a note at the end of his book, he stated, as regards the measurements of time, that "the whole subject has fallen into confusion;" that it was necessary to fudge in 3m. 3·68s. of purely imaginary time between the end of the equinoctial year of 1833 and the beginning of 1834, in order to "cook" theories with facts.

Considering that it is claimed by certain theorists that the rate of change in a star's right ascension is known to the one-thousandths of a second, it is rather startling to find that no less than 183·68 seconds had to be fudged in to make facts and theories agree.

How great was the confusion that existed not even Sir John Herschel suspected, but when the fact is stated that hitherto no observer, theorist, or mathematician has known in what a daily rotation of the earth really consisted, it can be understood that there has been, and will be, a confusion, until this problem is known and is dealt with in a correct manner.

The subject of time and its correct measurement is so vast, that only one item connected therewith can be dealt with in this book, but it can be stated, for the information of the reader, that there is connected with the measurement of time a problem hitherto unsolved, and which a geometrician who can free his mind from dogmatic theories will probably be able to solve. This problem will not be now dealt with, therefore it is open to any person for investigation and solution. The key to the solution may probably be found in the explanation which is given in this chapter, as to what constitutes a rotation of the earth.

The time used by observers consists principally of two kinds, viz. siderial time and mean solar time. The former is the time which will be mainly dealt with in this chapter.

A siderial day has been defined as the interval of time which elapses between two successive transits of the same star across a given meridian, and this day is divided into twenty-four hours.

An apparent solar day in the interval of time between two successive transits of the sun across the same meridian. The reason why sun time varies from sidereal time will be understood from an examination of the following diagram (Fig. 65).

The circle E F G H J represents the annual course of the earth round the sun, all details as regards the eccentricity of the orbit and its elliptical form being for the present omitted, as not necessary in the explanation which will now be given, of the difference between sun time and sidereal or star time.

The earth moves round the sun in the direction from E to F, G, H, and J.

We will suppose that on March 21 the earth is at E, and the sun S and a star at an infinite distance pass the meridian at the same instant of time.

On the next day the earth would have been carried in its orbit to F, and as the star referred to is at an infinite distance, a line drawn from F to the star would practically be parallel to a line drawn from E to the same star, but a line drawn from F to the sun would not be parallel to a line drawn from E to the sun. Consequently, the meridian

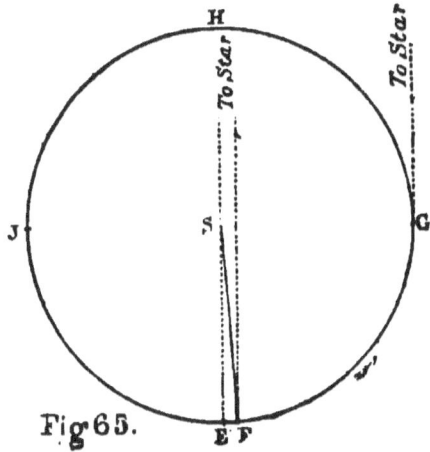

Fig 65.

would turn round by the daily rotation and come to the distant star before it turned over the small arc which brought this meridian to the sun. Hence the interval of time between two successive transits of the star would be less than the interval between two successive transits of the sun.

When the earth has been carried round one-fourth of its annual orbit to G, a line from G to the distant star would practically be parallel to the line drawn from E to the same distant star, but a line from G to the sun S forms a right angle (90°) with a line from G to the distant star.

The daily rotation of the earth being from right to left,

as represented in this diagram, the star will come on the meridian 90° before the sun when the earth is at G.

Now, as 24 hours of rotation represent 360° of daily rotation, 90° represent one-fourth, that is, six hours. The distant star, therefore, has gained six hours on the sun, owing to the movement of the earth from E to G.

When the earth has been carried to H, a line from H to the distant star will be parallel to the lines drawn from E or G to the distant star, and the earth at H is between the sun and the star.

The star, consequently, will come to the meridian 180° = 12 hours before the sun. The star, consequently, has gained half a day on the sun, in consequence of the movement of the earth from E to H.

For the same reason, the star will gain another interval of twelve hours in consequence of the movement of the earth from H to J and on to E.

It follows, therefore, as a geometrical law, that there must be one more transit of a distant star than there can be of the sun during the movement of the earth round the sun.

If the earth rotated only one hundred times during its journey round the sun, there would be a hundred transits of a star, but only ninety-nine transits of the sun.

The earth, in round numbers (for minute accuracy is not yet necessary for explanation), rotates yearly 365·25 times as regards the sun, but 366·25 times as regards a distant star.

There being one sidereal day more during a year than there are mean solar days, we can, by simple arithmetic, find the proportion between solar and sidereal time as follows:—

Twenty-four hours equal 1440 minutes. Divide 1440 by 365·25, and we obtain 3m. 56·55s. Consequently, 24 hours of mean solar time equal 24h. 3m. 56·55s. of sidereal time.

Again, divide 1440 minutes by 366·25, and we obtain 3m. 55·906s., which, subtracted from 24 hours, gives 23h. 56m. 4·09s. for the mean solar time, equivalent to 24 hours of sidereal time.

It will be evident, from an investigation of the last diagram, that the interval between two successive transits of the sun does not give the true time of the earth's rotation round its axis. It gives the time of one rotation plus the small arc over which the earth must rotate in order to bring the sun again on the meridian.

The interval between two successive transits of a star gives more nearly the measure of the time occupied by the earth in rotating, because the movement of the earth round the sun, owing to the immense distance of the stars, does not produce any apparent change in their relative positions.

The difference between apparent and mean solar time is explained (after a fashion) in most works on astronomy. This subject will not be dealt with in this chapter, but untrodden ground will be passed over, this being an investigation of what constitutes *a rotation of the earth.*

Sir John Herschel and his numerous copyists assert, " All the *stars*, it is true, occupy the same interval of time between their successive appulses to the meridian or to any vertical circle " (see Article 143, " Outlines of Astronomy ").

Do they indeed? Geometry teaches quite a different thing. Geometry teaches that a star within the circle traced by the earth's axis, will come to the meridian once oftener during an entire revolution of the equinoxes, or during one complete circle in the heavens described by the earth's axis, than will a star outside this circle.

Let this question be decided by facts, and it must be remembered that, no matter whether the time be long or short, during which a movement occurs, the laws as regards this movement are equally sound.

Suppose O P Q R (Fig. 66) the circle which the pole of the heavens describes among the fixed stars; C, the centre of this circle, the radius being of any length.

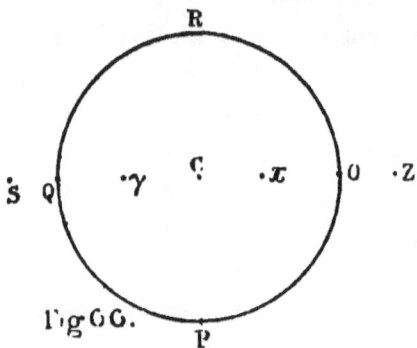

Fig 66.

Take O as the position of the pole at some remote date in the past.

The stars x, C, γ, and S would transit a meridian simultaneously, whilst the star Z would transit a meridian twelve hours before these stars.

After a long interval, the pole is carried to P.

When the pole is at P, the star Z will transit a meridian before the star x by an interval of time measured by the time it occupies for the earth to rotate through the angle Z P x.

The star x, consequently, instead of coming to the meridian twelve hours after the star Z, as was the case when the pole was at O, comes to the meridian after the star Z by an interval of time measured by the time it occupies the earth to rotate through the angle Z P x.

The stars C, γ, and S will come to the meridian after the star x by an interval of time measured by the time it occupies the earth to rotate through the angles x P C, x P γ, and x P S.

MEASUREMENT OF TIME, AND RIGHT ASCENSION. 199

When the pole has reached the point Q, then the stars γ, C, x, and Z will transit a meridian simultaneously, and each of these stars will therefore have gained twelve hours, or half a sidereal day, on the star Z. The star S, however, will, when the pole is at Q, transit twelve hours after the star Z, just as it did when the pole was at O.

When the pole has reached R, the stars γ, C, and x will transit before the star Z by an interval of time measured by the time it occupies the earth to rotate through the angles γ R Z, C R Z, and x R Z.

When the pole has again reached the point O, the stars x, C, and γ will transit a meridian twelve hours *before* the star Z instead of twelve hours after the star Z. Consequently, if θ represents the number of transits of Z, a star outside the circle described by the pole during any number of thousands of years, θ + 1 will represent the number of thousands of transits of a star within the circle described by the pole.

The interval between one or one million transits of the star x will not give the same measure of time as will the interval between one or one million transits of the star γ. There is a considerable amount of daily rotation to be performed after the star x is on the meridian and the pole is at P, before the star γ comes on the meridian; whereas when the pole is at Q, the stars γ and x will transit simultaneously.

How, then, can an authority assert that each star occupies exactly the same interval of time to come to the meridian; or, in other words, that the interval of time between two successive transits of any star gives the same measure of time?

The fact really is, that no two stars give exactly the same interval of time for their successive transits, and it is a geometrical law that a star within the circle described by the earth's axis will transit a meridian once oftener during the tracing of this circle than will a star which is outside this circle.

But we have another remarkable assertion promulgated by so-termed authorities, viz. that the time occupied by the earth in performing a sidereal rotation is to be measured by the interval between two successive transits of the same star.

Which star? Is it a star within, or without, the circle described by the pole? Is it a star near the circle described by the pole, or is it a star many degrees distant from this circle?

As it is a geometrical law, that a star within the circle described by the pole transits a meridian once oftener than a star outside this circle, it is absolutely necessary that, before we can assert that a star has a proper motion, we must know whether this star is within or without the circle described by the pole; we must also know how many degrees the star is from this circle.

Is the radius of this circle 23° 27', 24°, or of some other value? Even according to orthodox theories, this radius was formerly 24°, but is now only about 23° 27'.

How about the star Vega (a Lyræ)? Is this star within, or without, the circle described by the pole? If without the circle, how many degrees distant is it from this circle?

Will one million transits of the star a Lyræ give the same measure of time for the earth's rotation as one million transits of the pole-star (a Ursæ Minoris)? Certainly not, even if the present accepted theories were correct.

With these facts before us, what are we to think of the scientific knowledge of those persons who have asserted that "all the stars occupy the same interval of time between their successive appulses to the meridian," and that the actual time of a rotation of the earth is to be thus accurately defined.

The above is only one example of the slovenly manner in which this problem of time has been dealt with. Here is another.

There is a movement of the earth occurring which causes the earth's axis of daily rotation to change its direction. Hitherto theorists have been contented to accept, as a full and complete explanation of this movement, the assertion that it was "a conical movement of the earth's axis."

Let us examine the following diagram, which represents the earth, the actual instrument by aid of which we measure time (Fig. 67):—

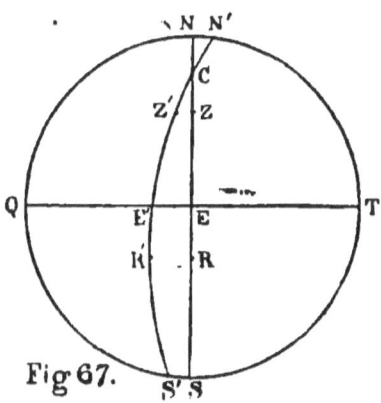

Fig 67.

Let N Q S T represent the earth; N the north, S the south pole; Q E T, the equator; N E S, a meridian of eighteen hours right ascension.

We will first assume that no other movement takes place in the earth than its daily rotation round the axis N S. At the completion of 366 rotations of the earth, the meridian N E S would again coincide with N E S.

But another movement does occur in the earth. During these 366 rotations, the pole N is carried to N', and the pole S is carried to S'.

202 UNTRODDEN GROUND IN ASTRONOMY AND GEOLOGY.

In what manner will the meridian N' S' be affected by this other movement, which, although independent of, is yet mixed up with, the daily rotation?

In what manner will the zeniths of localities, such as C, Z, E, and R, be affected by that movement which causes N to be carried to N', and S to S'?

The reader who has understood the earlier chapters in this book, knows that the point C remains fixed as regards this movement, that Z is carried to Z', E to E', and R to R'. Consequently, after 366 rotations of the earth, the meridian which, during the first rotation, occupied the position N C Z E R S will occupy the position N' C Z' E' R' S', each of these points except C being displaced by the second rotation. The movement of Z to Z', E to E', and R to R' is in opposition to the daily rotation, consequently the daily rotation has to occur slightly to bring the point E' back to E. Hence we should have 366 daily rotations completed when the meridian occupied the position N' C Z' E'. But it would require 366 rotations plus the small arc of rotation represented by E' E to bring the meridian to the same point on the equinoctial that this meridian occupied at the time of the first rotation.

Between the point C, however, and the pole N, a zenith x would be carried to x' in the same direction in which the pole N is carried, and the arc x x' is *in the same direction* as the daily rotation carries this zenith.

Hence the second rotation retards the daily rotation from C down to S', but accelerates the daily rotation from C to N'. We find, consequently, that a star between C and N—such, for example, as δ Ursæ Minoris—varied its right ascension annually,—in 1887 as much as $-$ 19·437s.; whereas another star, λ Draconis, varied its right ascension only $+$ 1·392s., the latter star being south of the point C, and the *daily* rotation affecting the readjustment of the locality under the star λ Draconis more rapidly than it does under the star δ Ursæ Minoris.

Can it, however, be asserted that the interval of time between two successive transits of these two stars gives the same measure of time, when, as it is proved by their annual changes in right ascension, there is for 366 transits a difference of more than twenty seconds?

By which of these stars are we to measure the time it occupies the earth to rotate? By which of the other stars are we to measure the time that it occupies the earth to rotate?

It is difficult to find any two stars which give the same interval of time for their successive transits when long periods of time are dealt with, and this result must follow as a geometrical law, due to the movement of the pole of the heavens in a circle round a point as a centre, which point has been hitherto in an unknown position in the heavens.

Every star within the circle traced by the pole of the heavens transits a meridian once oftener, during the time it occupies the pole to trace this circle, than does a star outside this circle.

Stars both within and without the circle traced by the pole will vary their relative rates of transit across a meridian.

There is but one point in the heavens in each hemisphere which will give a uniform standard measure of time, the successive transits of which point will give the exact measure of the earth's rotation on its axis; that point is the centre of the circle which the earth's axis traces, or, in other words, the pole of the axis of second rotation.

All stars between this pole of second rotation and the pole of daily rotation will give by their successive transits an interval of time less than that given by the successive transits of the pole of second rotation, hence the annual variation in right ascension of all stars so situated will in almost every case be a minus quantity, increasing in amount the nearer the star is to the pole of daily rotation. As examples of this change, the following stars may be referred to for 1887:—

Star.	Annual variation in right ascension.
ϵ Ursæ Minoris	— 6·3433s.
δ Ursæ Minoris	— 19·4370s.
γ Ursæ Minoris	— 64·4299s.

Stars which are south of the point C will all increase their right ascensions annually, but not uniformly, because those stars near the pole of second rotation, where the zenith is but slightly affected by this second rotation, will have their right ascension less affected than will those stars farther from the pole of second rotation. As examples of such effects, the following stars may be given for 1887:—

Star.	Distance from pole of second rotation.	Variation in right ascension.
β Draconis	9° 17′ 38″	+ 1·3515s.
γ Draconis	9° 6′ 22·5″	+ 1·3923s.

As examples of stars which most increase their annual right ascension, we must look for stars near the equator of second rotation, viz. stars which have somewhere near eighteen hours right ascension, and a south declination of near 29° 25′ 47″. As examples of such stars we find the following for 1887:—

Star.	South declination.	Right ascension.	Annual variation.
θ Ophiuchi	24° 53′	17h. 15m.	+ 3·6778s.
μ Sagittarii	21° 5′	18h. 7m.	+ 3·5838s.
α Sagittarii	25° 29′	18h. 21m.	+ 3·7020s.

Stars in this part of the heavens, but north of the equator of daily rotation, will be at a greater distance from the equator of *slow* rotation, and consequently will not be so much affected by the second or slow rotation as will those stars that are south of the equator of daily rotation, and consequently are nearer the equator of second rotation. The following stars will serve as examples :—

Star.	North declination.	Right ascension.	Annual variation.
α Ophiuchi	12° 38'	17h. 49m.	2·7792s.
β Ophiuchi	4° 36'	17h. 37m.	2·9598s.
μ Herculis	27° 47'	17h. 42m.	2·3440s.
α Lyræ	38° 40'	18h. 33m.	2·0304s.

It will be seen, by an examination of the various declinations of these stars, that the nearer a star happens to be to the equator of second rotation the greater will be its annual variation in right ascension, and the nearer a star approaches the pole of second rotation the less, consequently, will be its annual change in right ascension. At the pole of second rotation there will be little or no annual change in right ascension, but north of this pole of second rotation the annual change of right ascension will alter from plus to minus.

There will be another condition under which a star will vary its annual right ascension by a very minute quantity— that is, where an arc from the star to the pole of second rotation is nearly at right angles to the arc from the star to the pole of daily rotation.

Under these conditions, that part of the meridian over which the star is situated is carried towards or away from the pole of daily rotation, and the position of this part of the meridian is but slightly altered as regards the daily rotation.

For example (Fig. 68), take N C S as a meridian of eighteen hours right ascension; C, the pole of the axis of second rotation.

N Z R S, a meridian short of eighteen hours—say, for example, of seventeen hours right ascension. There will be a point in this meridian, such as Z, where an arc drawn from Z to C will be at right angles to an arc drawn from Z to N. This is the same thing as stating that an arc drawn from C to Z is at right angles to that part of the meridian on which Z is located.

Whilst the second rotation carries the pole N to N' round C as a centre, it carries Z to Z' round C as a centre. The arc Z Z' coincides with the meridian N Z; consequently a star which is situated above Z will have its right ascension altered very little, if any, by the second rotation.

A point γ on the same meridian would be carried to γ' round C as a centre by the second rotation; this star, consequently, which was in the zenith of γ, will have increased its right ascension by just the amount of *daily* rotation required to bring γ' to the meridian N' Z γ.

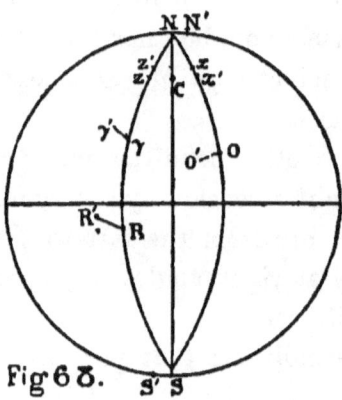

Fig 68.

As examples of this simple law, two stars are given below; their right ascensions, omitting seconds, are given, and their north polar distances, and the rate at which they varied their right ascensions annually at the date 1850, as found by observation.

Star.	North polar distance.	Right ascension.	Variation.
22 Draconis ζ	24° 6'	17h. 8m.	+ 0·159s.
65 Herculis δ	64° 58'	17h. 8m.	+ 2·459s.

The first-named star occupies a position near Z on the last diagram, where its annual variation in right ascension is very slight. The second star occupies a position near γ, where the zenith is considerably affected by the second rotation, and where, consequently, the annual variation in right ascension is considerable.

By taking a point farther down this meridian, such for example as R, we shall find that R R', the arc of displacement produced by the second rotation, is greater than $\gamma\gamma'$, because R is at a greater distance from C, the pole of second rotation; consequently, a star in the zenith of R would have a greater rate for its annual variation in right ascension than would the star above γ. The star 40 Ophiuchi ξ, 110° 56′ from the north pole, and with a right ascension in 1850 of about 17h. 12m., serves as an example, the annual variation in right ascension at that date being recorded as 3·592s.

The same law holds true as regards a meridian such as N x O S (Fig. 68), which we will suppose a meridian of 19h. 30m. right ascension.

A star near the zenith of x will have its right ascension altered very slightly annually, because this zenith x will be carried to x' round C as a centre, the arc $x\,x'$ nearly coinciding with the meridian N x O S.

That part of the meridian which is at O will be carried to O' round C as a centre; consequently, a star at the zenith of O will show a large variation annually in right ascension.

The two stars given below serve as examples of this law:

Star.	North polar distance.	Right ascension.	Variation.
58 Draconis π	24° 34′	19h. 19m.	+ 0·332s.
a Vulpeculæ	65° 38′	19h. 22m.	+ 2·495s.

The reader who comprehends geometry can now make an interesting calculation. He can calculate at what point on each meridian a star will very slightly change its right ascension annually. He can then trace out a curve or irregular figure joining these points. All stars with few exceptions within this curve will decrease their right ascensions annually. All stars outside the curve will increase their right ascensions annually.

The method of calculating this curve is very simple, and is as follows.

The angular distance between the pole of daily rotation P and the pole of second rotation C, viz. 29° 25' 47", is the hypotenuse of a right-angled spherical triangle, the right angle being on any particular meridian, say seventeen hours of right ascension.

The pole of second rotation is assigned a right ascension of eighteen hours. Hence for a meridian of seventeen hours we have the following spherical triangle, P C D:—

P C = 29° 25' 47".

P D C, a right angle.

The angle D P C = 1h. = 15°.

Calculate P D, the polar distance of that point on a meridian of seventeen hours where an arc from the pole of second rotation is at right angles to this meridian.

Using the common formula, we have—

Log. tangent, 29° 25' 47" = 9·7513982
Log cosine, 15° 0' 0" = 9·9849438
_____
9·7363420 = log. tan., 28° 35' 14" = P D.

For a meridian of 16h. substitute 2h. = 30° for 1h. and we obtain 26° 2' 20" for the polar distance of the point at which an arc from the pole of second rotation cuts the meridian of 16h. at right angles.

For a meridian of 15h. we obtain 21° 44' 53" for the polar distance of this point. For a meridian of 14h. we obtain 12° 37' 45"; for 13h. 8° 18' 26"; for 12h. 30m., 4° 12' 41"; for 12h., 0.

For 19h., the point will be 28° 35' 14" from P; for 20h., 26° 2' 20"; for 21h., 21° 44' 53", and so on.

MEASUREMENT OF TIME, AND RIGHT ASCENSION.

Hence we can construct this curve as follows (Fig. 69):—
Describe a semicircle to represent that half of the sphere from twelve to twenty-four hours right ascension.

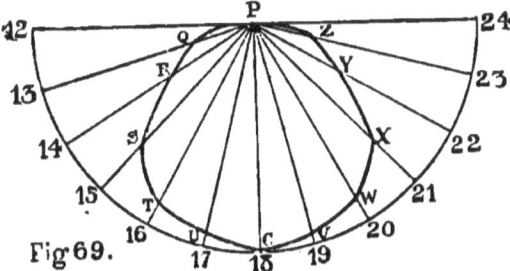

Fig. 69.

Take P as the north pole of daily rotation; C, the pole of second rotation; P C consequently = 29° 25′ 47″.

Set off the various meridians at 15° interval = 1h. The numbers 12, 13, 14, etc., represent meridians of twelve hours, thirteen hours, fourteen hours right ascension.

From the calculations already given, we can mark on each meridian that point at which an arc from C will cut each meridian at right angles, thus:

$$P Q = 8° 18' 26'' = P Z$$
$$P R = 12° 37' 45'' = P Y$$
$$P S = 21° 44' 53'' = P X$$
$$P T = 26° 2' 20'' = P W$$
$$P U = 28° 35' 14'' = P V$$
$$P C = 29° 25' 47''$$

The following are the geometrical laws appertaining to this curve.

All stars outside of this curve will increase their right ascensions annually, the rate of increase being least the nearer these stars are to the curve; this law holding true from twelve to twenty-four hours right ascension.

Stars within this curve will as a rule decrease their right ascensions, except under conditions named below. The nearer a star is to the pole of daily rotation, the greater will be the annual rate of *decrease* in the right ascension of this star.

P

Whilst every point on this curve will be carried directly towards that point where the pole was located, it may not be carried directly towards that point where the pole at the end of one year will be located. The pole P is carried annually over an arc of 20·09″ in the direction of twenty-four hours right ascension (nearly). A point a few degrees within the curve at eighteen hours right ascension will be carried annually over a small arc round C as a centre far less than 20·09″. Consequently, the meridian may, under certain conditions, cause stars near this curve, although within it, to increase very slightly, instead of decreasing, their right ascension annually.

The calculations of these details are too long and too technical to be here elaborated. Some few examples, however, of the general effects may be given.

The star β Ursæ Minoris is near the meridian marked P S, having about 14h. 51m. right ascension. The distance of this star from the pole P is about 15° 15′, whereas P S is more than 21°. Hence the annual rate of variation in this star will be *minus*.

The star 4949 (British Association Catalogue) Draconis, which has a right ascension of about 14h. 55m., and is distant from the pole P about 23° 28′, will have an annual variation in its right ascension plus in value, because outside the curve.

Another star, viz 4982 (British Association Catalogue) Ursæ Minoris, with about fifteen hours right ascension, but only about 6° 53′ from the pole P, will have a large rate of decrease in its annual right ascension.

Below are given the various annual rates of change in right ascension of these stars for the date 1850, as found by observation:

β Ursæ Minoris	− 0·273s.
4949 Draconis	+ 0·935s.
4982 Ursæ Minoris	− 4·797s.

MEASUREMENT OF TIME, AND RIGHT ASCENSION. 211

A star, such for example as β Bootis, about 49° 1' from the pole P, increases its right ascension annually about 2·264s., and this star has a right ascension of about 14h. 56m.

It might be interesting to inquire whether a true rotation of the earth is to be measured by the interval between two successive transits of β Bootis, or of 4982 Ursæ Minoris. Taking even 366 rotations of the earth, there would be a difference of 7s. in this rate, according as we elected to take one or the other star.

It is not true, therefore, that the daily rotation of the earth can be measured by the interval of time which elapses between two successive transits of any star, or that all stars by their successive transits give the same interval of time. If a star be selected, this star may give a longer or shorter interval of time by two successive transits than the earth occupies in completing one rotation.

Here, then, we have one cause for that confusion to which Sir John Herschel candidly called attention; but there are other causes, which have yet to be made known.

One of the most important elementary matters to be first dealt with is to separate what is a true rotation of the earth round its axis of daily rotation, and then what is the effect of the second rotation in retarding this daily rotation in some parts, but accelerating it in other parts.

The daily rotation will be accelerated between the poles of second rotation and of daily rotation, but retarded between the poles of second rotation and other points of the earth's surface.

The equator of second rotation cuts a meridian of six hours right ascension at a point in latitude (terrestrial) 29° 25' 47" north. This point on the earth's surface has its daily rotation retarded annually by the second rotation by an arc of 40·9". The same locality referred to a meridian of eighteen hours right ascension will be twice 29° 25' 47" =

58° 51′ 34″ from the equator of second rotation, and will be retarded by the second rotation only by an arc of 40·9″ × cosine of 58° 51′ 34″ = 21·149″ annually.

These two arcs, being referred to the equator of daily rotation, will be increased as follows:—

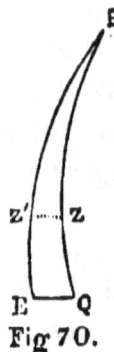
Fig 70.

$$Z\ Z' = 40·9''$$
$$Q\ Z = 29°\ 25'\ 47''$$
$$\text{Therefore } E\ Q = \frac{40·9''}{\text{Cos. } 29°\ 25'\ 47''} = 46·9''$$
Again, if $Z\ Z' = 21·149''$
$$E\ Q = \frac{21·149''}{\text{Cos. of } 29°\ 25'\ 47''} = 24·2''$$

An arc of the equator of 46·9″ converted into time, as indicated by the daily rotation equals 3·1s.; an arc of the equator of 24·2″ converted into time equals 1·6s. Hence the daily rotation required to readjust, as it were, the effects of the second rotation for a meridian of six hours right ascension amounts to 3·1s. annually, whilst the daily rotation required to readjust the effects of the second rotation for a meridian of eighteen hours right ascension amounts to 1·6s. annually. Hence there is a difference of 3·1s. − 1·6s. = 1·5s. in the apparent effect of the daily rotation during a year between the meridians of six hours and eighteen hours right ascension, as regards a latitude of 29° 25′ 47″ north.

It follows that the earth would have to perform 1·5s. more of its daily rotation to bring a star with six hours right ascension on the meridian, than it would have to perform to bring a star on the meridian which had eighteen hours right ascension, these stars in each case having a north declination of 29° 25′ 47″.

Although there are no two stars situated exactly in the positions referred to, we yet find that there are two sufficiently near to the points to serve as examples. These stars are as shown below, their right ascensions, declinations, and annual variation in right ascension being extracted from a star catalogue of 1850.

Star.	Declination north.	Right ascension.	Annual variation.
κ Aurigæ	29° 32' 51"	6h. 5m. 49·14s.	+ 3·828s.
o Herculis	28° 44' 43"	18h. 1m. 41·53s.	+ 2·341s.
		Difference	1·487s.

From this example, a geometrician may perceive how each case may be worked out, and the simple calculation can be made. Take, for instance, a star with six hours right ascension, and 10° north declination; this star would be 19° 25' 47" from the equator of second rotation. Then take a star with eighteen hours right ascension, and 10° north declination; this star will be 39° 25' 47" from the equator of second rotation. Work out the effects of the second rotation on these two stars, and it will be found that the annual variation in right ascension of the star having six hours right ascension will exceed that of the star having eighteen hours right ascension by about 0·5s.

Again, take two stars on these meridians, with 10° *south* declination, and the opposite effect would be manifested, viz. the star with six hours right ascension would be 39° 25' 47" from the equator of second rotation, whereas the star with eighteen hours right ascension would be 19° 25' 47" from the equator of second rotation. Hence the annual variation of the star with six hours right ascension would be *less* by about 0·5s than would that of the star with eighteen hours right ascension, but with the same declination. Although no stars are situated exactly in the positions to serve as examples of this law, the two stars given below give from their positions the correct results, viz.—

Star.	Declination south.	Right ascension.	Rate.
θ Leporis	14° 55'	5h. 59m.	+ 2·718s.
6279 Sagittarii	14° 39'	18h. 20m.	+ 3·419s.
		Difference	0·701s.

When the effects of the second rotation come to be examined as regards meridians of twelve and twenty-four hours right ascension, none of those hitherto mysterious variations occur which take place in connection with six and eighteen hours right ascension. The reason why these variations do not occur is very simple, and can be readily understood after an examination of the following diagram:—

P E S Q (Fig 71) represent the earth; P the north, S the south, pole; E Y Q, the equator of daily rotation; P O Z X Y S, a meridian of twelve hours right ascension; C, the position of the pole of axis of second rotation.

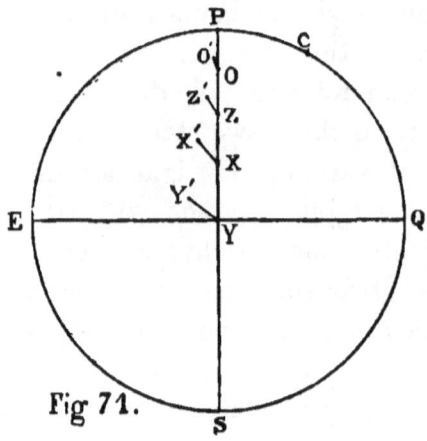

Fig 71.

The various points O, Z, X, and Y are carried by the second rotation annually over small arcs round C as a centre, consequently O is carried to O', Z to Z', X to X', and Y to Y'. But these changes are very similar to those which are produced by the *daily* rotation failing to complete one rotation. Let the daily rotation be completed, and O', Z', etc., would by the *daily* rotation be brought almost exactly into the meridian P, S. Their zeniths, in each case giving a difference in zenith distances of stars, but no great *variation* in the annual rate of right ascension.

The same law holds good for a meridian of twenty-four

MEASUREMENT OF TIME, AND RIGHT ASCENSION. 215

hours right ascension, each zenith being carried south over small arcs round C as a centre.

Hence, when we examine the annual variation in right ascension of a star within some 20° or 30° of the pole, and of one some 80° or 90° of the pole, we find but a very slight difference in the annual rate of change in right ascension. Here are some examples:

Star.	North polar distance.	Right ascension.	Annual rate.
δ Ursæ Minoris	32° 8'	12h. 8m.	3·016s.
γ Corvi	106° 42'	12h. 8m.	3·077s.
6 Comæ	74° 16'	12h. 8m.	3·054s.
2 Canum Venaticor.	48° 30'	12h. 8m.	3·034s.
η Virginis	89° 50'	12h. 12m.	3·067s.

When we turn to twenty-four hours right ascension, we find that the same uniformity in rate must occur in the annual variation in right ascension, stars very near the pole of daily rotation being affected by those changes already described in connection with the last curve given; for example (1850).

B.A.C. Star.	North polar distance.	Right ascension.	Annual variation.
8334 Cassiopeæ	29° 37'	23h. 54m.	+ 3·009s.
29 Piscium	93° 52'	23h. 54m.	+ 3·075s.
22 Andromedæ	41° 46'	0h. 2m.	+ 3·093s.
γ Pegasi	75° 39'	0h. 5m.	+ 3·084s.
α Andromedæ	61° 44'	0h. 0m. 38s.	+ 3·086s.

When we examine the results of the second rotation as affecting the annual rate of variation in right ascension of two stars having nearly the same north declination, the one star being near twenty-four hours, the other being near twelve hours, right ascension, we shall not find those great changes or differences in this rate which occurs with stars having six and eighteen hours right ascension. The reason for this is evident. The zeniths of twenty-four hours and twelve hours right ascension are nearly at the same distance from the equator of slow rotation, no matter what the latitude of each locality may be, as long as these latitudes are in each case the same; but with six hours and eighteen hours the same latitudes are at very different distances from the equator of second rotation, and are consequently very differently affected. Hence we find that two stars having nearly the same declination or polar distance, but the one being near a meridian of twenty-four hours right ascension, the other near a meridian of twelve hours right ascension, will vary their relative annual rate very slightly. Take for example, the two stars given below for 1850:

Star.	North polar distance.	Right ascension.	Annual variation.
β Cassiopeæ	31° 40′ 39″	0h. 1m. 12s.	+ 3·149s.
δ Ursæ Minoris	32° 8′ 1″	12h. 7m. 58s.	+ 3·016s.

Now, if these two stars had six and eighteen hours right ascension, the effect of the second rotation would cause their annual rates to vary as much as 1·5s.

Again, take the two stars given below, viz.:

Star.	North polar distance.	Right ascension.	Annual variation.
γ Pegasi	75° 40′	0h. 5m. 31s.	+ 3·084s.
4125 Comæ	74° 16′	12h. 8m. 22s.	+ 3·054s.

Here we have a difference in the annual rate of only three-hundredths of a second per year. When, however, we examine the difference in rates for stars with about six and eighteen hours right ascension, the results are very different. Example—

Star.	North polar distance.	Right ascension.	Annual rate.
ν Orionis	75° 13′ 7″	5h. 59m. 0s.	+ 3·428s.
6049 Herculis	73° 14′ 12″	17h. 53m. 22s.	+ 2·667s.

Here we have a difference of seventy-six-hundredths of a second in the rate.

The cause of these variations will be evident from an examination of the diagrams in this chapter.

It can be at once seen that the second rotation does not produce similar results on the two meridians of six and eighteen hours right ascension, but it does produce similar effects on twelve and twenty-four hours right ascension.

Another and most important result of the second rotation is, that a locality 29° 25′ 47″ from the equator of

MEASUREMENT OF TIME, AND RIGHT ASCENSION. 217

daily rotation is retarded annually by the second rotation by an arc of daily rotation amounting to 40·9″ for this latitude, when referred to a meridian of six hours right ascension, and this arc of 40·9″ is directly in opposition to the daily rotation.

This same locality is retarded by the second rotation over an arc less than 40·9″ when referred to twelve hours right ascension, and to be found as follows:—

$$C\ P = 29° 25' 47''$$
$$P\ Z = 90° - 29° 25' 47'' = 60° 34' 13''$$
Angle $C\ P\ Z = 90°$
Therefore $C\ Z$, the hypotenuse $= 64° 39' 45''$

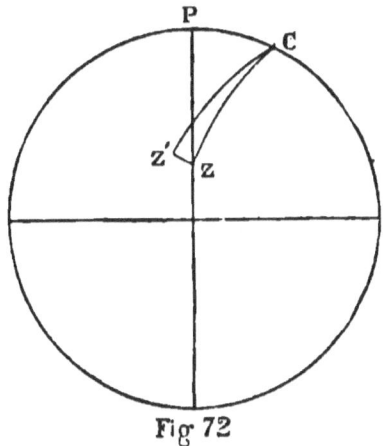

Fig 72

The point Z, therefore, is 25° 20′ 15″ from the equator of second rotation, and the length of the arc Z Z′, over which the second rotation carries this point, is 40·9″ × cosine of 25° 20′ 15″ = 36·9″.

But this arc Z Z′ is not in *direct* opposition to the daily rotation; it is only partially in opposition. The daily rotation will carry the point Z′ to the meridian P Z over a small arc, found as follows:—

$$Z\ x = 20\ 09''$$
$$Z\ Z' = 36·9''$$
Angle at $x = 90°$
Calculate $Z'\ x$

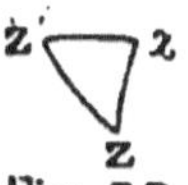

Fig 73.

$Z'x$ will be about 30″.

Hence, whilst the daily rotation has to readjust the point on the meridian of six hours by a rotation over an arc of 40·9″, it has for twelve hours to perform a rotation of only 30″ to reach the same meridian it occupied the year previously.

Here we have an interesting fact on which geometricians may exercise their skill, inasmuch as this result of the second rotation has something to do with the assumed variation in the eccentricity of the earth's orbit round the sun, inasmuch as the zenith of such a locality as that named appears to separate annually from six hours to twelve hours by a small arc.

We may now gather together the principal facts connected with the measurement of time and of right ascensions, and note what these prove.

First, it is not true, as has been stated, that "all the stars occupy the same interval of time between their successive appulses to the meridian." A star within the circle described by the pole of the heavens will transit once oftener whilst this circle is being described than will a star outside the circle. Hence it is necessary, for this purpose alone, to know exactly the radius of the circle which the pole does describe.

Secondly, the time occupied by the earth in making one rotation can be measured by one item only, viz. by the successive transits of that point in the heavens which is unaffected by the second rotation, and which consequently is the pole of the axis of second rotation.

Thirdly, the measurement of sidereal time from that point on the equinoctial termed the first point of Aries is a cause of confusion, inasmuch as this point shifts annually, and does not shift uniformly; nor will it shift uniformly until the date 2295·2 A.D. Neither does this

MEASUREMENT OF TIME, AND RIGHT ASCENSION. 219

point shift to an equal amount that the line of the solstices shifts, for the reason given in the latter part of this chapter.

Fourthly, the differences that occur in the relative rates for the annual variation in right ascension of stars in meridians of six hours and eighteen hours right ascension, and which differences do not occur with stars on twelve and twenty-four hours right ascension, are due to the geometrical laws which follow the effects of the second rotation.

Fifthly, the hitherto unaccountable changes in stars' right ascension which take place between meridians of fifteen hours and twenty-one hours, are due to the effects of the second rotation, as already explained.

These changes are not due to the assertion that the whole solar system is rushing towards the constellation Herculis at the rate of one hundred and fifty-four million miles per year, and that sooner or later a general crash will occur, and the whole universe will be destroyed. Such remarks are promulgated by those persons who, either from ignorance or laziness, are not acquainted with the laws of geometry and mechanics.

CHAPTER XIV.

MODERN ASTRONOMICAL OBSERVATIONS.

There is probably no performance so likely to impress a visitor, or even a board of visitors, with the supposed perfection of modern astronomy as to attend at a large observatory, and to witness some important observation being made.

The huge transit instrument, looking much like a piece of heavy artillery, revolves noiselessly on its pivots, and could be moved by even a child. The various adjustments of the instrument can be readily made; and, as far as the machinery is concerned, everything is perfect. Even the observer is reduced to a sort of machine, his personal error of observation having by repeated trials been ascertained and consequently allowed for.

When the meridian transit of the sun occurs, various members of the staff occupy positions which enable them to read the micrometers, by which hundredths and thousandths of a second are ascertained. By long practice the observers become so expert that, although great haste is used, there is no apparent hurry; and when it is announced that the results of this transit have been arrived at to the ten-thousandth of a second, a visitor would probably exclaim, "Wonderful!" and he would no doubt be convinced that modern astronomy was now in such a state of perfection, that to question the minute accuracy of any portion of it was a proof of ignorance only.

It must be admitted that instrumental observation, as far as it goes, is very near perfection; but there is a serious item to be considered in connection with astronomical science, and even as regards this instrumental observation.

More than seventeen hundred years ago there was a very able observer named Ptolemy, who used, for determining the position of the celestial bodies, various instruments. These instruments were of three different kinds. One was a meridional armillæ, consisting of two rings, with which he measured the zenith distance of the sun and the obliquity of the ecliptic. The exterior of these two rings was graduated in degrees, each degree subdivided to as small an amount as was possible.

Another instrument was a quadrant with a very large radius, and was used to measure zenith distances.

The third instrument has been termed "the parallactic rods," which consisted of a pillar placed vertically on a base. At the upper extremity was a joint, on which turned a long alidade carrying two sights. This instrument was also used to determine zenith distances.

Another instrument also used by Ptolemy was an astrolabe, used for measuring the distance between the sun and the moon, or between the moon and a star.

In order that observations should be made under all conditions, the ancients dug caves, the entrance to the cave being opposite the meridian. Thus Strabo relates ("Geogr.," lib. xvii.) that in his time, at Heliopolis, caves were shown which had been used by observers for ascertaining the position of celestial bodies by day. There were also at Cnidus, in Asia, similar caves used for the same purpose.

All these arrangements, and the vast instruments used at the time, were no doubt as perfect as the mechanical skill at that date could accomplish. They have been surpassed in accuracy by the moderns, it is true, but they accomplished with moderate accuracy only that which at the present day is accomplished with great accuracy, viz ascertaining by observation the positions of the celestial bodies. The ancient observers could foretell with very fair accuracy when an eclipse would occur, when a star would come to the meridian, when a star would be occulted by the moon, etc., etc.; and yet these observers believed that the earth was stationary, and neither rotated every twenty-four hours on its axis, nor revolved annually round the sun.

If a visitor, or a board of visitors, had gone to the observatory of Ptolemy, had examined his grand instruments, seen how he made his observations, and verified the fact that he could predict the day and hour at which an eclipse or an occultation of a star would occur, these persons would undoubtedly have asserted that the science of astronomy was perfect, and that any person who doubted this perfection must be very ignorant. Yet two important items were unknown at that date, viz. the daily rotation and annual revolution of the earth round the sun.

It is a fact, therefore, that in spite of the accuracy which may be arrived at by perpetual observation as regards the predicting the positions of celestial bodies for a few years in advance, yet the causes which produce the movements may be unknown.

It follows, therefore, that although the instruments and observations of the ancients were admirable, considering the means at their disposal, and the instruments and observations at the present time are very near perfection, yet in each case it is a question of observation only, and does not give any proof that the true movement of the earth is known at the present day any more than it was seventeen hundred years ago.

It may appear a strange assertion to those persons who

take everything for granted, yet it is a fact, that if the earth did not move, whilst the various celestial bodies revolved daily round it, every item now obtained by observation could be equally as well obtained by the same means, and a Nautical Almanac could be framed on exactly the same lines as one is now framed.

Hence, although instrumental observations may be carried out with great perfection, we must not make the mistake of imagining that *therefore* the science on which these observations are made is perfect. The ancients did make this mistake, and consequently during more than fourteen hundred years their system of astronomy was wrong, and they considered it a duty to ignore or persecute all those who, being gifted with reasoning powers, asserted that it was more probable that the earth rotated each twenty-four hours, than that the heavens revolved round the earth during the same period of time.

It will probably be admitted that, had instruments been, in the days of Sir Isaac Newton, in as perfect a condition as they now are, this philosopher would soon have made himself acquainted with their mechanical details, and could, as easily as any modern observer, have read by aid of his instruments the supposed zenith distance of a star to the one-thousandth of a second. What would have been the result? A visitor might be struck and impressed with the most wonderful accuracy thus obtained, and would probably imagine that perfection had been reached. Unfortunately, however, Sir Isaac Newton framed a table of refractions for various altitudes, and he imagined that in summer the refraction for certain altitudes varied from that in winter as follows:—

Altitudes.	Summer.	Winter.
15°	0° 3' 4"	0° 3' 28"
20°	0° 2' 17"	0° 2' 33"
25°	0° 1' 46"	0° 2' 0"
30°	0° 1' 26"	0° 1' 36"

Hence it would follow that, whilst he imagined he was by his instruments giving results accurate to the one-thousandth of a second, he was wrong more than 31″ for an altitude of 15° measured in summer, the refraction for 15° being now taken as 0° 3′ 35″.

The effect produced by refraction varies according to the state of the atmosphere, and must ever bring an element of uncertainty to bear on observations; it is not too much to admit that this uncertainty may amount to 1″. Consequently, whilst the instruments used for observations may give readings to one-thousandth of a second, refraction may cause these readings to give results erroneous to 1″. The minute accuracy, therefore, which is claimed by observers is theoretical only, as it cannot exist as long as refraction affects the position of a star as seen with a telescope.

But does the minute accuracy which is claimed actually exist?

For example, in the Nautical Almanac for 1873, the mean right ascension for the star δ Ursæ Minoris for January 0 − ·226d. is given as 18h. 13m. 18·497s. In the Nautical Almanac for 1887, the mean right ascension for the same star for January 0· + ·165d. is given as 18h. 8m. 45·996s. Consequently, between these dates the mean right ascension of this star has decreased 4m. 32·501s. = 272·501s. during fourteen years—that is, at the mean rate of − 19·4643s. per year. Yet in the Nautical Almanac for 1873, it is stated that the annual variation in right ascension for this star is − 19·3924s., and in 1887 it is − 19·4370s.

These discrepancies occur in a multitude of instances, proving that the minute accuracy claimed is theoretical only. It is, of course, very easy to explain such matters by asserting that the star itself moves; but the fact remains that, whilst it is claimed that accuracy exists to the fourth place of decimals, facts prove that even the second place of decimals does not agree.

Is it, however, probable that accuracy to the ten-thousandth of a second can be practically arrived at, when, as stated by Sir John Herschel, it was found necessary to add more than three minutes of purely imaginary time, in order to cook the accounts?

The statement of Sir John Herschel, that the whole subject of "time" is in confusion, is corroborated by facts, and it will remain in confusion until theorists become more practical, and really investigate the true movement of the earth as a geometrical problem. As long as authorities are contented to accept as highly satisfactory the vague definition that the earth's axis, by the joint action of the sun and moon, has "*a* conical movement," they will be compelled to depend entirely on continued instrumental observation for their results. When, however, they really investigate the second rotation of the earth, they will discover that observations are a clumsy and inaccurate method of arriving at results.

Among the numerous "visitors" and other persons connected with observatories, it is a most remarkable fact that these individuals seem to be fully convinced that astronomy is in a most perfect condition, because very fairly accurate observations can be made. If reason were brought to bear on this question, the inquiry ought to be, Why are observations any longer necessary? No instrument that man ever made can be so accurate and uniform in its movements as is the earth itself. Some changes may occur at long intervals of time, but these changes must be very slow, and one or two observations per year would be sufficient to check any irregularity, and reveal in what direction this was occurring. Yet we find that the mere routine observer usually claims it as a proof of the importance of the work carried out at his observatory, that upwards of one hundred observations per year have been made of various important stars, not at one observatory only, but at scores of observatories.

What is the use or object of all this labour, when every item can be calculated to within a fraction of a second for fifty or a hundred years, when the second rotation of the earth is comprehended? It would be a somewhat puerile proceeding to employ a number of persons to count how many times the second-hand of a chronometer rotated during each minute, hour, and day of the year, and to claim as most important the labours of individuals so employed. A chronometer correctly constructed will cause the second-hand to rotate sixty times during each minute, 3600 times during each hour, and 86,400 times during each day of twenty-four hours.

The polar distance and right ascension of each star can be calculated with equal facility. If we want to know what the mean polar distance of the pole-star, for example, was or will be for any date past or future, the formula is given in Chapter X. If we want to know what the apparent polar distance of this star will be for any day of the year, we can measure this by aid of the nautilus curve given in Chapter XI. To claim, therefore, that it is necessary to make some hundred observations annually of any star, is a proof, not of the efficiency of astronomy, but of the incompetency of observers to arrive at the results by any other means.

It seems, also, little short of marvellous, that although from the time of Ptolemy, on to that of Bradley, and down to the present date, instruments have been used to measure meridian zenith distances, yet it has never been considered worth while to investigate in what manner various zeniths were affected by "a conical movement of the earth's axis."

Observers have carefully noted how the zenith of the pole of daily rotation has been affected annually by this movement; but how other zeniths were affected did not seem to trouble them. Observation, perpetual observation alone, was the end of everything; good instruments, a large staff of highly paid observers, night after night devoted to the transit instrument and chronometer, day after day devoted to correcting the observations, and astronomy was claimed to be in the most perfect condition, in spite of the fact that the mean polar distance of not one single star could be calculated by any sound geometrical law for even ten years in advance or in the past, the only method known being to approximate to this position by means of adding or subtracting some annual rate, found by repeated observation only.

That instrumental observations can now be made with great accuracy, is no more a proof of the perfection of astronomy than it was in the time of Ptolemy. In fact, the necessity for maintaining a large staff, whose duty it is to continue night after night making these observations, is a proof that there is something not yet known connected with the movement of the earth. We have not far to search in order to find what this something really is. When we find that a vague movement of the earth termed merely a conical movement of the axis is accepted as satisfactory although no mention is made of which pole remains fixed, and which gyrates, or how each zenith is affected by this conical movement, it is easy to understand why perpetual observation is assumed to be the only available means by which star-catalogues for the future can be framed.

The reader may now comprehend how great will be the difference in the real value of instrumental observation when the second rotation of the earth, with all its details, becomes understood. Hitherto it has been by means of these observations only, that the position which a star would occupy at a date only four or five years in advance could be assigned. Now, however, there is a means by which the position of this star can be calculated for a hundred years or more, without reference to more than one observation. Instead, therefore, of being dependent on instrumental observation, we can actually check these, and can decide whether instruments have been in adjustment, observers have been careful, and whether the correct refraction has been used at various dates.

Take, for example, any star, say β Ursæ Minoris, the mean right ascension of which star for 1887 is—

$$\left. \begin{array}{l} \text{January } 0^\circ + \cdot 165\text{d.} = 14\text{h. } 51\text{m. } 2\cdot 486\text{s.} \\ \text{The mean declination} = 74^\circ\ 37'\ 1\cdot 88'' \end{array} \right\} \text{recorded observation.}$$

From this data alone, calculate the mean declination of this star for any other dates—say January 1, 1850, 1819, 1755, and 1950—and show what errors, if any, were made by observers at former dates.

To the mere routine observer, this calculation could no more be accomplished than a Zulu could work out a quadratic equation. A geometrician who understands the second rotation of the earth can solve the problem without difficulty. The solution has been already given herein, and in "Thirty Thousand Years of the Earth's Past History" (Chapman and Hall).

CHAPTER XV.

THE PLANE AND THE POLES OF THE ECLIPTIC.

THE plane of the ecliptic is the course along which the earth travels annually round the sun. It may be defined as that circle in the heavens which the earth would trace annually among the fixed stars if an observer were located in the sun. It is also the course which the sun's centre appears to trace among the fixed stars when seen by an observer on the earth's surface.

The poles of the ecliptic are two points in the heavens, distant 90° from all parts of the ecliptic. These poles bear to the ecliptic the same relation which the poles of the earth bear to the equator, or the poles of the heavens bear to the equinoctial.

Theoretically, it is a very simple problem to determine exactly the position which the plane, and hence the poles, of the ecliptic occupy in the heavens. Practically, it is one which presents difficulties which, it appears, are so great that since astronomy has been a science it has never been accomplished.

The four earliest catalogues of stars with which we are acquainted are those formed by Ptolemy, 140 A.D.; Ulugh Beigh, 1463 A.D.; Tycho Brahè; and Hevelius. In these catalogues the latitudes of stars are given, that is, their least angular distance from the plane of the ecliptic, and their longitudes, which is the angular distance or arc of the ecliptic intercepted between two circles of latitude, one passing through the star, the other passing through the vernal equinox.

It must be remembered that, no matter what minor instruments were used by the ancients, yet they employed for their observations the one great instrument which the moderns use, viz. the earth itself; and it was as necessary in ancient days as it is at present, that the exact movements of this great instrument should be correctly known.

The determination of the position of the pole of the ecliptic by the ancients appears to have been conducted in a very vague manner. Ptolemy, for example, gave the position of this pole as 24° from the pole-star, 14° 20' from β Draconis, and 7° from δ Draconis. When this position is assigned, it differs from that which Ptolemy gave it by three other stars, for he assigned its position as 14° 30' from γ Draconis, 24° 30' from a Draconis, and 15° 20' from θ Draconis.

Tycho Brahè gave the pole of the ecliptic 23° 58' from the pole-star, thus indicating that from the time of Ptolemy to that of Tycho Brahè there had been by these observations a decrease of only 2' in the distance of the pole of the ecliptic from the pole-star, although during the same period there had been a decrease of nearly 30' in the distance of the pole of the heavens from the pole of the ecliptic.

Upon checking the various latitudes of stars as assigned by various ancient observers, and hence attempting to fix the position which they imagined the pole of the ecliptic occupied, we encounter a confusion only, leading to the conclusion that they could not with accuracy fix the position of this pole.

When we come to modern times, we find equally as

interesting a subject for reflection. The modern observer starts from a theoretical assumption. He assumes that the course which the pole of the heavens traces on the sphere of the heavens is a circle round the pole of the ecliptic as a centre. He concludes that when the pole in its course neither increases nor decreases its distance from a star, that therefore the pole of the ecliptic must be on the arc joining that star and the pole of daily rotation, and at such a distance from the pole of daily rotation as the obliquity of the ecliptic happens to be at the time.

This conclusion might be true, supposing that during the past thousand or more years there had not been the slightest variation in the obliquity of the ecliptic, because there being no change, would prove that the pole of the heavens did not vary its distance from the pole of the ecliptic, and hence that the pole of the heavens must trace *a circle* round the pole of the ecliptic *as a centre*. When, however, as is known, the obliquity decreases, and has decreased during two thousand years at least, then it is impossible that the pole of the heavens can trace *a circle* round the pole of the ecliptic *as a centre*, and the method adopted to determine the position of the pole of the ecliptic is erroneous.

In order to demonstrate the very serious error which both the ancient and modern observers have made in this matter, the following diagram (Fig. 74) can be examined.

The circle N O P Q R is the circle traced by the pole of the heavens during one second rotation round C as a centre, the radius C O = C P = C Q = 29° 25′ 47″.

E is the position of the pole of the ecliptic; C E = 6°. R is the position which the pole of the heavens will reach at the date 2295·2 A.D.

When the pole of daily rotation at a remote date in the past was at O, a star, x on the meridian joining O and C, would neither increase nor decrease its polar distance; that is, for a brief period O would not vary its distance from *x*. The modern theorist, therefore, would assume that the pole of the ecliptic must be on the arc O *x* C; and, as the angular distance O E would be the measure of the obliquity at that date, he would place the pole of the ecliptic at A, the distance O A being equal to O E.

The pole of the heavens, continuing its uniform movement round C as a centre, would after many years reach P, at which date a star *y* on the arc joining P and C would not increase or decrease its polar distance for a short period. The theorist would now assume that the pole of the ecliptic

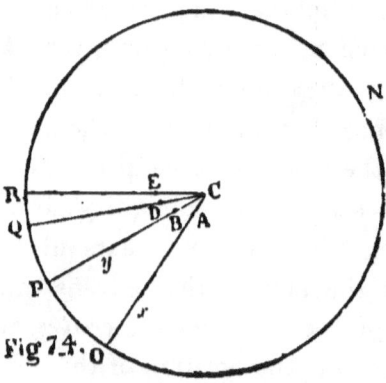

Fig 74.

was on the arc P *y* C, and distant from P as much as the obliquity was at that date.

P E would represent the obliquity at that date, consequently P B would be made equal to P E, and the pole of the ecliptic would be assumed to be at B, on the arc P C.

In the same manner, when the pole of the heavens was at Q, the pole of the ecliptic would be assumed to be at D, on the arc Q C, the distance Q D being made equal to Q E, the obliquity at that date.

We can now examine how, by this erroneous system, the pole of the ecliptic has been assumed to move. At a remote

THE PLANE AND THE POLES OF THE ECLIPTIC. 233

date it was assumed to be at A, then at B, then at D; and it was given two movements—one towards the pole of daily rotation, the other laterally as regards this pole.

This is exactly the movement which theorists imagine the pole of the ecliptic does make, and for the reasons given above.

Although it is admitted that it is a movement of the earth itself which causes the polar distances of stars to vary, and causes also the precession of the equinoctial points, *yet*, in consequence of a blind adherence to the theory that the pole of the heavens traced *a circle* round the pole of the ecliptic *as a centre* (in spite of the impossibility of this belief being true, as long as a decrease in the obliquity occurred), another movement had to be invented in order to explain facts, viz. a shifting of the pole of the ecliptic, with all the confusion and contradictions which follow.

One movement of the earth fully explains all the facts, and enables a geometrician to calculate with accuracy the obliquity for any date, as also the positions of various stars. The present accepted theory commences with an assumed movement which is a geometrical impossibility, and then invents another movement, viz. that of the pole of the ecliptic, in order to endeavour to escape from the impossibility.

The stars themselves, however, refuse to obey the present accepted theory; but their evidence is easily disposed of by theorists. All those stars which disprove the assumed movement are accused of having an enormous amount of proper motion; then, arguing in circles, it is asserted that it is proved that certain stars must have a very large amount of proper motion, because they do not agree with accepted theories.

In vol. xxx., page 329, of the *Philosophical Trans-*

actions, there is a paper by Edmund Halley entitled, "Considerations on the Change of the Latitudes of Some of the Principal Fixed Stars."

After some remarks on the rate of the precession, he continues: "But while I was on this inquiry, I was surprised to find the latitudes of three of the principal stars in heaven directly to contradict the supposed greater obliquity of the ecliptic, which seems confirmed by the latitudes of most of the rest; they being set down in the old catalogue [Ptolemy's] as if the plane of the earth's orbit had changed its situation among the fixed stars about 20′ since the time of Hipparchus. Particularly, all the stars in Gemini are set down, those to the northward of the ecliptic, with so much less latitude than we find, and those to the southward with so much more southerly latitude. Yet the three stars Palilicium, or the Bull's-eye, Sirius, and Arcturus contradict this rule directly; for by it Palilicium, being in the days of Hipparchus in about 10° of Taurus, ought to be about 15′ more southerly than at present; and Sirius, being then in about 15° of Gemini, ought to be 20′ more southerly than now; yet, on the contrary, Ptolemy places the first 20′ and the other 22′ more northerly in latitude than we now find them. Nor are these errors of transcription, but are proved to be right by their declinations set down by Ptolemy as observed by Timocharis, Hipparchus, and himself, which show that those latitudes are the same as those authors intended. As to Arcturus, he is too near the equinoctial colure to argue from him concerning the change in the obliquity of the ecliptic; but Ptolemy gives him 33′ more north latitude than he has now, and that greater latitude is likewise confirmed by the declinations delivered by the said observers. So, then, all these three stars are found to be above half a degree more southerly at this time than the ancients reckoned them. When, on the contrary, at the same time, the bright shoulder of Orion has in Ptolemy almost a degree more southerly latitude than at present. What shall we say, then? It is scarcely credible that the ancients could be deceived in so plain a matter, three observers confirming each other."

Halley committed the same error that his predecessors and followers have committed. He imagined that no variation in the obliquity of the ecliptic could occur, except by a change in the position of the plane of the ecliptic. He ignored the movements of the earth entirely, as though these had nothing to do with the problem, and seemed to overlook the fact that, if the theory were true that the pole of the heavens always traced *a circle* round the pole of the ecliptic *as a centre*, these two poles must always maintain the same distance from each other, no matter how much the ecliptic altered its position; and, consequently, no change whatever could occur in the obliquity if the pole of the heavens did always maintain the same distance from the pole of the ecliptic, as it must do if it traced a circle round this pole as a centre.

It would appear something little short of marvellous to find men, during hundreds of years, failing to perceive these elementary errors, and ignoring the movement of the earth in connection with the possible variation in the obliquity of the ecliptic. When, however, we are reminded how, during many hundreds of years, authority after authority adopted the follow-my-leader system, and denied the daily rotation of the earth as an absurdity, whilst the Epicycles of Ptolemy were accepted as profound and truly scientific, the apparently marvellous confusion receives a simple explanation.

It would have occurred to most reasoners whose minds were not slaves to authority and routine, that, as the earth was known to have some movement which produced a change in direction of its axis, possibly this movement had not been quite correctly interpreted, and that this same movement might explain, when correctly read, the decrease in the obliquity. The solution was very simple; for if the pole of the ecliptic happened not to be the centre of the circle which the earth's axis traced, the problem was at once solved. But it had been asserted that the earth's axis did trace a circle round the pole of the ecliptic *as a centre*, and never varied its distance from this centre, in spite of the admitted fact that it did, and had decreased its distance from this centre during the past two thousand years at least. Hence it was supposed that if the plane, and hence the pole, of the ecliptic was made to vary its position, a solution of this contradiction would be given, in spite of the geometrical law which proves that as long as the pole of the heavens maintains a uniform distance from the pole of the ecliptic, no change in the position of either pole can produce any change in the obliquity.

Why the movement of the earth should have been so utterly ignored in this investigation, whilst the possible movements of the plane of the ecliptic should have received so much attention, exhibits a remarkable phase of the theoretical mind. More especially is this remarkable when it is a geometrical law that, if the theory were true that the pole of the heavens did trace *a circle* round the pole of the ecliptic as *a centre*, then no variation in the position of the pole of the ecliptic could produce any variation in the obliquity, because the angular distance of these two poles is the exact measure of the obliquity.

Reference may now be made to some of those singular results which follow the present system of counting right ascensions, and determining the earth's position in its orbit.

THE PLANE AND THE POLES OF THE ECLIPTIC, 237

The initial point from which observers count right ascensions is the first point of Aries; that is, the point on the ecliptic which the sun's centre occupies at the period of the vernal equinox.

In consequence of the change in direction of the earth's axis produced by the second rotation, amounting to about 20·09" annually, the pole of daily rotation, which was 90° from the sun's centre at the vernal equinox of one year, will be 90° − 20·09" from the sun's centre on reaching the same point in its orbit on the year following. Hence the equinox will occur on this following year before the earth has completed its circle round the sun, the small arc by which it fails to complete its circle being ascertained in the following manner :—

$v\ x$ (Fig. 75) represents a portion of the ecliptic; v E, a portion of the equator, or equinoctial. x E = 20·09", the amount of the movement of the pole of daily rotation produced by the second rotation. The angle $x\ v$ E represents the obliquity of the ecliptic for the year for which we wish to find the precession. Let us take the obliquity as 23° 27' 14", which it was about 1887. To calculate $v\ x$, we use the usual formula for a right-angled spherical triangle v E x, and the hypotenuse $v\ x$ is about 50".

Fig 75.

Consequently, the equinox each year falls short of its position of the year previous by about 50" and a small decimal. The reader, however, must not commit the error of imagining that this 50" is a constant quantity for all time, although the second rotation occurs uniformly. The value is dependent on the obliquity or angle $x\ v$ E; the greater the obliquity, the less the annual precession. It would be a serious error to make, knowing that the obliquity is a variable quantity, to assert that because now, when the obliquity is about 23° 27', and the precession consequently about 50" per year, or about 1° in 72 years, therefore a whole revolution of the equinoctial points would occur in 360 times 72 years—that is, during 25,920 years. This assertion has been promulgated by theorists, and is incorrect, as any geometrician can perceive who examines the problem. The annual precession varies from about 51" to about 34", depending on the value of the obliquity. For example, when the obliquity was 35° 25' 47", and the annual movement of the pole of daily rotation 20·09", we should find $v\,x$, last diagram, = 34".

The equinox of one year precedes the equinox of the previous year at the present time by an arc of the ecliptic equal to about 50". Consequently, the initial point from which right ascensions are counted shifts along the ecliptic 50" annually. The earth, consequently, has to rotate this arc of 50" to bring the same point on the meridian, and as 1° of daily rotation occupies 4m. of time, 50" occupies about 3·3s. of time, an amount by which the right ascension of stars are, under normal conditions, increased annually.

It must now be borne in mind that the arc of the ecliptic intercepted between two successive vernal equinoxes amounts to about 50". An examination will now be made of that which occurs at the winter solstice.

The winter solstice occurs when the sun's centre and the pole, or imaginary pole, of the ecliptic transit a meridian simultaneously. At this instant the polar distance of the sun's centre will be at a maximum, and the sun's south declination consequently at a maximum.

In consequence of theorists having imagined that the pole of the ecliptic shifts its position, so as always to be on that meridian where the stars for the time being do not vary their polar distances, we have the following results (Fig. 76).

C represents the pole of second rotation; P, the position

of the pole of daily rotation; L, a point on the equator of slow rotation; L P C being a meridian on which the stars do not vary their polar distances. The meridian L P C would therefore be a portion of the solsticial colure.

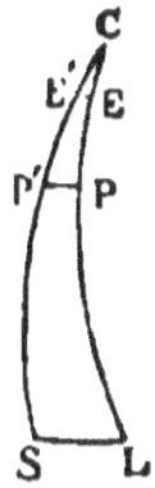

The pole of the ecliptic would by theorists be placed at E on this arc, and distant from P as many degrees, minutes, and seconds as the obliquity happened to be for that year.

Fig 76.

One year afterwards, the pole P would have been carried to P' round C as a centre, the radius CP = CP' = 29° 25' 47"; the arc P P' = 20·09", and the arc S L consequently being 40·09".

The meridian S P' C would now be that meridian on which stars did not vary their polar distance, and which consequently would be a portion of the solsticial colure.

The pole of the ecliptic would now be placed at E' on the meridian S P' C, the angular distance P' E' being made equal to the obliquity for that year.

We can now calculate how much the pole of the ecliptic is made to shift annually by this assumption.

The angular distance C E = 29° 25' 47" − P E, the obliquity. Let us take this obliquity as 23° 27' 14", in round numbers; then C E = 5° 58' 33", and L E = 84° 1' 27", consequently E E' = 40·9" × cosine of 84° 1' 27" = 4·2".

Thus the pole of the ecliptic is assumed to have a lateral movement of about 4·2", and to approach the pole P' about 0·46" per annum.

Each year the pole of the ecliptic would be imagined to move in a similar manner, in order to suit accepted theories.

It appears somewhat singular that, whilst a multitude of theories have been evolved from the imagination of speculators as regards astronomical matters, very few of which have any foundation in fact, it never seems to have occurred to these individuals to produce the various arcs from the pole of daily rotation through the imaginary positions of the pole of the ecliptic; had they done so they might have discovered that all these arcs intersected at the same point in the heavens, viz. at C, the pole of the second axis of rotation. Instead, therefore, of having the course of the pole of daily rotation defined as it is at present, viz. *a circle* described round a movable centre, the true nature of this curve would have been defined, viz. a circle round a point which is not coincident with the pole of the ecliptic.

Whilst, however, two vernal equinoxes intercept between them an arc of the ecliptic which at the present time is about 50" per year, two winter solstices intercept between them an arc of the ecliptic of only $40·9'' \times$ cosine of $84° = 40·59''$.

Whilst, then, the vernal equinox has an annual precession at the present time of slightly more than 50", the winter solstice has an annual precession of only 40·59". Hence the interval of time between the winter solstice and the vernal equinox gradually decreases, from a cause quite independent of the eccentricity of the earth's orbit round the sun. If the earth's orbit were a perfect circle, and the sun were in the centre of this circle, the second rotation of the earth would produce this effect at the present time.

As long as the annual precession of the equinoctial point is greater than the annual rate of the second rotation, viz. 40·9" per annum, the vernal equinox will decrease its distance from the winter solstice.

When the obliquity was 29° 39', the annual precession of the equinoctial point was about 40·9" only, or nearly of the same value as the precession of the winter solstice at the present date.

(241)

CHAPTER XVI.

SOME EFFECTS OF THE SECOND ROTATION.

SOME of the minor effects of the second rotation of the earth have been explained in the preceding pages, but many others have yet to be brought into notice. The reader who has made himself acquainted with the details given in the previous chapters will know that he possesses a means of arriving by calculation at results with the greatest accuracy, which results have hitherto been entirely beyond the power of mathematicians and theorists. All the pomp and circumstance of large observatories, huge transit instruments, a staff of highly paid observers and computers, and the hundreds of observations made annually of each star, are to the person acquainted with the details of the second rotation of the earth merely laborious and expensive proceedings, adopted by those individuals who know no other and more accurate method of arriving at the same results.

This fact there is no disputing. Numerous examples have been given in this book, and hundreds more can be given if necessary, to prove that from one accurate observation only, the mean polar distance of a star can be calculated for one hundred years or more to within 1″, without any reference to the annual rate of change in polar distance of this star, found by the present system by observation only. The reader should test this fact for himself; he will then

R

probably estimate at a proper value the mere opinions of theorists who assert that such results cannot be calculated.

Here is another example which the reader may test for himself—whether the polar distance of a star can be calculated by aid of a knowledge of the second rotation of the earth, and by means hitherto unknown.

THE STAR β CORVI (Fig. 77).

P C = 29° 25′ 47″
C β = 106° 19′ 15·2″ } Constants.

Angle P C β, January 1, 1887 = 107° 33′ 52″.
Annual variation in angle P C β = 40·9″.

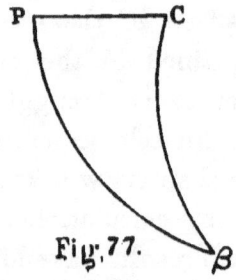

Fig. 77.

The 40·9″ to be subtracted for each year in the past.

From this data calculate P β, the mean polar distance of this star for any date in the past, and find how near to accuracy observers were at past dates.

The two following values are extracted from star catalogues at two distant dates for the polar distance of this star:—

January 1, 1850 ... ... 112° 34′ 0·1″
    „   1780 ... ... 112° 10′ 38″

Compare calculation with these records, or with any others available.

We next come to the precession of the equinoctial point, which can be calculated by aid of a knowledge of the second rotation of the earth, as proved in Chapter V., and we perceive how great a mistake theorists have made in asserting that the period of an entire revolution of the equinoxes could be obtained by merely multiplying by

360, the number of years during which 1° of precession occurred.

Perhaps the most remarkable oversight, however, committed by theorists, is that of M. La Place and his numerous unreasoning copyists, viz. that no variation can or ever could occur in the obliquity of the ecliptic, except by a variation in the plane of the ecliptic, which is the course of the earth round the sun.

That a considerable variation must occur in the obliquity, if the earth's axis traced a circle round any other point in the heavens as a centre than the pole of the ecliptic, seems never to have occurred to these individuals. The possible movement of the earth itself was entirely overlooked, and it was asserted that the only possible manner in which any variation in the obliquity could occur was by a variation in the position of the plane of the ecliptic.

The most astounding inconsistency, however, still remained as an accepted dogma, viz. that the pole of the heavens always traced a circle round the pole of the ecliptic as a centre. If this movement did occur, then it mattered not how much the plane of the ecliptic varied, because no variation in this plane could by any possibility produce the slightest change in the obliquity. It is not possible to find in even ancient astronomy any greater muddle, contradiction or absurdity, than exist in connection with these assertions of the modern theorist.

The reader acquainted with the second rotation of the earth will perceive that the statement that the pole of the heavens traces a circle round the pole of the ecliptic as a centre, is a gratuitous and impossible assumption, contradicted by the well-known fact that the pole of the heavens has, during the past two thousand years at least, gradually decreased its distance from this assumed centre. He must be aware that, whilst the modern theorist and observer know that the distance of the pole of the heavens from the pole of the ecliptic varies some 46″ per century at the present date, yet the cause of this decrease is unknown. It is supposed that this decrease of 46″ per century is due to some movement in the plane of the ecliptic; but no movement in this plane could produce any change in the obliquity if the earth's axis moved, as has been asserted, in a circle round the pole of the ecliptic as a centre.

The reader may ignore all these contradictions, and can, by aid of a knowledge of the second rotation, calculate the value of the obliquity for thousands of years, just as easily as he can calculate the polar distance of a star for a hundred years, independent of any observations. He can demonstrate why the rate in the decrease in the obliquity is now about 0·46″ per year. He can calculate what the obliquity was one thousand or ten thousand years ago, and need not trouble himself about "solar tables" any more than he need be dependent on the Epicycles of Ptolemy. Here is the formula by which he can calculate the obliquity of the ecliptic for any date, and the rate for any year independent of observation, tables, or theories:

$$(2295 \cdot 2 - T) \times 40 \cdot 9'' = < C.$$

T being the year for which the obliquity is required.

$$C E = 6°$$
$$C P = 29° 25' 47''$$

With two sides and the included angle at C calculate P E the obliquity.

Examples: calculate the obliquity for January 1, 1800, without any reference to solar tables or annual rates, etc. For the date 1800 the following is the calculation: $(2295 \cdot 2 - 1800) \times 40 \cdot 9'' = 5° 37' 33 \cdot 68''$; with the two sides C E = 6°, C P = 29° 25' 47″, and the included angle at C = 5° 37′ 33·68″, calculate P E, the obliquity for 1800

A.D. The result gives 23° 27' 55·6", which was within 1" of what observers found it at that date.

For any other dates the obliquity can be calculated with equal facility and accuracy, independent of any observation; and we now arrive at the fact that at the date about 5624 B.C. the second rotation of the earth caused the obliquity to amount to 29° 30', and at the date 13,544 B.C. to amount to 35° 25' 47".

What were the conditions as regards climate on earth at the date 13,544 B.C. may now be examined (Fig. 78.)

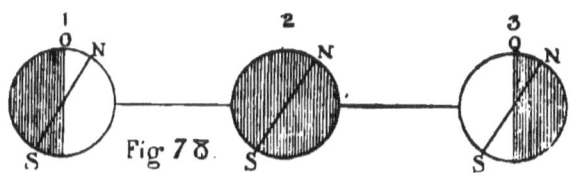

No. 1 represents the earth at the period of the summer solstice at the date 13,544 B.C.; N S, the position to which the earth's axis has been carried by the second rotation; O, a point on the earth's surface distant from N, the north pole, 35° 25' 47". No. 2 represents the position of the earth at the autumnal equinox; No. 3, the position of the earth at the winter solstice, N S being the axis of daily rotation, and Q a point on the earth's surface 29° 25' 47" from N. The conditions which exist at the present time relative to the earth and its orbit, will be understood from the following diagram :—

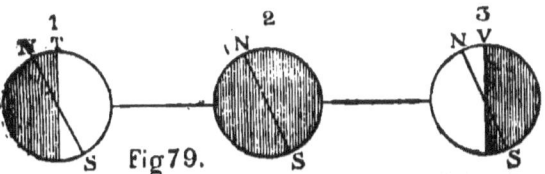

No. 1 represents the earth at the period of the winter solstice at the present date; N S, the axis of daily rotation; T, a point on the earth's surface 23° 27' from N. No. 2

represents the earth at the vernal equinox: No. 3, the earth at the summer solstice; N S, the axis of daily rotation.

The sun is in the plane represented by the line joining the centres of the circles at No. 1 and No. 3.

Under the present conditions, the arctic circle extends from N to T, No. 1, and N to V, No. 3; and, as a geometrical law, whatever may be the extent of the arctic circle, the tropics or points on the earth's surface at which the sun is vertical at midday extend to the same amount from the equator.

At the present time the arctic and antarctic circles extend about 23° 27' from the poles, and the tropics extend about 23° 27' from the equator.

At the date 13,544 B.C. the arctic and antarctic circles extended to a distance from the poles of 35° 25' 47", and the tropics extended to 35° 25' 47" from the equator. This condition was produced by that same second rotation of the earth by aid of a knowledge of which we can calculate the polar distance of a star by one observation only, and for a hundred years or more. It is merely carrying back that same movement to 13,544 B.C., and of the earth itself, that enables us to arrive at this result.

The plane of the ecliptic need have no movement; it is the earth which moves, not the orbit of the earth, round the sun.

The assertion that no change in the obliquity or in the extent of the arctic circle can possibly occur, except in consequence of a change in the plane of the earth's orbit round the sun, is a statement utterly untrue, and is opposed to the elementary laws of geometry. To continue repeating such an erroneous dogma exhibits a condition of the mind little short of imbecility.

We can now examine the climatic conditions which prevailed on earth at the date 13,544 B.C. during each year.

At the period of the vernal equinox in spring to the northern hemisphere, the sun was, as now, vertical at the equator, and the day was twelve hours long. As the earth travelled in its course round the sun, the days would increase in length, but more rapidly than they now increase. At the period of midsummer the sun would have 12° more altitude at midday than at present for each locality in the northern hemisphere; consequently, for 55° north latitude the sun's meridian altitude would be found as follows :—
The meridian altitude of the equinoctial for 55° latitude is 90° − 55° = 35°, and the sun, having 35° 25' 47" north declination, would have a meridian altitude of 35° + 35° 25' 47" = 70° 25' 47".

At midnight, however, at that date, and under such conditions, the sun, having a north declination of 35° 25' 47", would be only 90° − 35° 25' 47" from the north pole of the heavens; that is, 54° 34' 13" only from the north pole.

It is a geometrical law that the altitude of the pole at any locality is equal to the latitude of that locality, hence for 55° north latitude the altitude of the pole would be 55°. But the sun at midnight would have been only 54° 34' 13" from the pole, and would therefore have been visible as a midnight sun from 55° north latitude.

During several days, therefore, both before and after midsummer at 13,544 B.C. the sun would have remained above the horizon, having a midday altitude of 70° 25' for 55° north latitude, and just touching the horizon at midnight.

An increase in the midday altitude of the sun of 12° at that date would cause much greater heat in high northern regions than now exists, and icebergs and melted snow would be liberated from the arctic regions, and scattered down to lower latitudes.

On the autumnal equinox being reached, the sun would

be again vertical at the equator, and the days would, as at present, be twelve hours long.

At the winter solstice at 13,544 B.C. the sun would have 35° 25′ 47″ south declination, and as, for 55° north latitude, the meridian altitude of the equinoctial is only 35°, the sun would not be visible at this latitude, darkness or only twilight would prevail, and intense cold would occur, just as now occurs in arctic regions during winter. Every lake and river would be frozen, and the country covered by a mantle of snow. Scotland, Wales, and the north of England would have their glaciers, and all similar latitudes north and south would be similarly affected.

The following summer would melt either all or a portion only of these effects, but the scattering of boulders and drift materials would have been enormous each year.

The effects of the second rotation as it at present exists would have been most marked in middle latitudes, not in the vicinity of the equator or of the poles.

The migration of all animals at that date must have been very great, the variation of climate from summer to winter being far more extensive than at present. Arctic animals might come down to 51° latitude in winter, and tropical animals travel to 60° latitude or thereabouts in summer.

During many thousand years these conditions would prevail, because at the date 5624 B.C. the arctic circle extended to about 29° 30′, at 13,544 B.C. it extended to 35° 25′, and at 21,460 B.C. it reached to 29° 30′; there was, therefore, a period of 15,866 years, during which considerable alterations of climate between summer and winter prevailed in both hemispheres.

On tracing back the course which has been followed by the earth's axis on the sphere of the heavens during the past two thousand years, as proved by recorded observa-

SOME EFFECTS OF THE SECOND ROTATION. 249

tions, and by the accurate calculations which can be made by a knowledge thereof, it would appear as shown in the following diagram (Fig. 80) :—

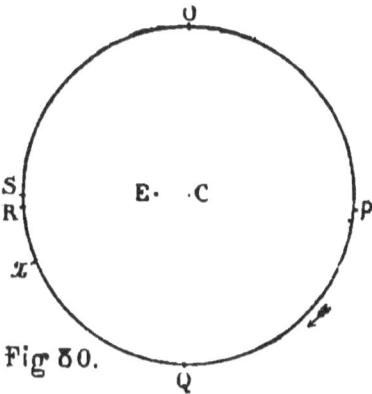

C is the pole of the second axis of rotation projected on the sphere of the heavens; E, the pole of the ecliptic; O P Q R S, the circle which the earth's axis traces on the sphere of the heavens during three-fourths of the second rotation. At 21,460 B.C. the pole was at O, when O E, the obliquity, was 29° 30'. P, the position of the pole at 13,544 B.C., at which date P E, the obliquity, was 35° 25' 47"; E Q, the obliquity at 5624 B.C., when it was 29° 30'; R E, the obliquity at 1887 A.D., when it was about 23° 27' 14"; S E, the obliquity at 2295·2 A.D., when it will be 23° 25' 47".

It will be seen from this diagram how gradually the great changes of climate came on from 21,460 B.C., until they reach a maximum at 13,544 B.C., from which date these great annual alternations decreased, until Q was reached at the date 5624 B.C. From Q the decrease was more rapid until a point x was reached at about 2000 B.C., when the decrease in the distance of the pole from E, the pole of the ecliptic, was also rapid. What was the decrease per century at that date can be easily calculated as follows:—

From 2295·2 A.D. to 2000 B.C. was 4295·2 years. The second rotation takes place at the rate of 40·9″ annually. Multiply 40·9″ by 4295·2, and we obtain 48° 47′ 45″ for the angle E C x for the date 2000 B.C. With the two sides, C E = 6° and C x = 29° 25′ 47″, and the included angle at C, the obliquity E x can be calculated, and will be found 25° 50′ 24″ for the date 2000 B.C.

Find the obliquity for the date 1900 B.C. By merely substituting 1900 for 2000 and proceeding as in the last example, the obliquity for 1900 B.C. will be found 25° 44′ 21″.

Between 2000 B.C. and 1900 B.C. the obliquity decreased 6′ 3″.

If a theorist at the date 1900 had asserted that, as the obliquity decreased 6′ 3″ since the date 2000 B.C., *therefore* it would decrease 1° 0′ 30″ for 1000 years, and *therefore* 4° 2′ 0″ during 4000 years, and consequently at the date 2000 A.D. the obliquity would be 4° 2′ 0″ less than it was at 2000 B.C. viz. 21° 48′ 24″, he would have shown that he was not only ignorant of the cause which produced the decrease in the obliquity, but was also unacquainted with the elementary law of geometry, which proves that a curve cannot decrease its distance from a point at a uniform rate.

The modern theorist has committed these errors; he has asserted that because the obliquity now decreases 46″ per century, it must have decreased 460″ during 1000 years, and 4600″ = 1° 16′ 40″ during 10,000 years.

We have, then, first the palpable error, that no change in the obliquity can possibly occur except by a change in the plane of the ecliptic, the movement of the earth being thus utterly ignored.

Secondly, there is the self-evident contradiction that the earth's axis traces a circle round the pole of the ecliptic as a centre, whilst it is admitted that these two poles have decreased their distance during 2000 years at least.

Thirdly, it is asserted that because the decrease in the obliquity now occurs at the rate of about 46" per century, *therefore* it has decreased at the same rate during the past 10,000 years.

Fourthly, the actual movement of the earth which causes the zeniths of the poles of daily rotation to move over arcs of 20·09" annually is left undefined, as though it was of no consequence how every other zenith moved.

Fifthly, whilst it is asserted that every movement of the earth is known, it is absolutely necessary to maintain at various observatories a large staff of observers, in order to obtain by repeated observation the rate of change in the polar distance and right ascension of stars, so that a catalogue of stars' positions can be framed for three or four years in advance. If the movements of the earth were really known, the results now obtainable by observation only, could, as is proved in this book, be arrived at by calculation.

Perhaps the most amusing proceeding of the theorist is, when he makes assertions which he imagines to be unanswerable, and seems to believe that he has finally disposed of any objections which have been brought forward against his pet theories.

Some years ago, when geologists and reasoners were convinced that the changes of climate shown by geology to have occurred in former times must be due to astronomical causes, M. La Place announced that any such changes were impossible, *because* the plane of the ecliptic could not vary more than 1° 21' either way.

The reader will perceive that this theorist ignored the movement of the earth when he made the above announcement. If, however, any person at that date had hinted at the possibility of the plane of the ecliptic varying as much as 2° instead of 1° 21', he would have been hooted out of every scientific society, for daring to question the infallibility of the French theorist.

Several years elapse, and another French theorist (M. Leverrier) announces that the plane of the ecliptic can vary only 4° 52' from a mean position. It follows, therefore, if this later theory be correct, then M. La Place was quite wrong in asserting that 1° 21' was the limit of the change in the obliquity.

Strange as it may appear to the geometrician or reasoner, the movements of the earth itself were entirely ignored in both these theories. It seemed to be imagined that if the earth's axis traced its circle or course round a point even 20° from the pole of the ecliptic, yet no change in the obliquity could occur except by means of, and due to, a change in the plane of the ecliptic itself.

This erroneous assumption was probably due to the theory that the earth's axis always traced *a circle* round the pole of the ecliptic *as a centre*, the fact being overlooked that, if the earth's axis did trace such a circle, no change whatever could occur in the obliquity, no matter how much the plane, and hence the pole, of the ecliptic changed their positions.

But we are again assured by theorists that this variation in the plane of the earth's orbit, the only supposed means by which they imagine a variation in the obliquity can occur, and which they state is limited to 4° 52' either way, is just at the present date in the mean condition, so that, taking 23° 28' as the mean, they assert that the obliquity, due solely for its variation on the changes in the plane of the ecliptic, will reach to 23° 28' + 4° 52' = 28° 20', and to 23° 28' − 4° 52' = 18° 36'. Yet during all this change of 9° 54' in the supposed centre, the pole of the heavens is imagined to trace *a circle* in the heavens round the pole of the ecliptic as a centre.

If, however, as appears to be claimed, the exact changes in the plane of the ecliptic and the course traced by the earth's axis be so accurately known, why is it necessary to demand tens of thousands of pounds annually, in order to pay observers and computers to find out how the pole of the heavens is moving? Why cannot these simple items be calculated?

It is proved in this book that such items can be calculated, for a hundred years at least, when the second rotation of the earth is known, and without reference to more than one observation.

Why is it claimed, as a proof of the valuable and important work carried on at an observatory, that some hundred observations per year have been made of several hundred stars?

Is it assumed that these stars will suddenly start off and run away somewhere unless they are continually watched?

When the true courses of the pole and the zenith are known, we can foretell for a hundred years exactly the polar distance and right ascension of these stars, and it is proved in this book that such results can be obtained.

When the true courses of the pole and zenith are not known, and when the meridian changes in a manner that is not comprehended by theorists, then, and then only, are those perpetual observations necessary which are now practised by observers.

In order to fit the contradictions relative to the movement of the earth's axis with orthodox theories, it has been asserted that, no matter how much or in what manner the pole of the ecliptic moves, yet the earth's axis always traces a circle round the pole of the ecliptic as a centre. This movement is as impossible as is the former, and is devoid of any foundation in fact. Let the geometrician produce various arcs drawn from the pole of daily rotation through those points to which theorists shift the pole of the ecliptic from time to time, and they will find that all these arcs intersect at the same point in the heavens, viz. at the pole of the second axis of rotation, 29° 25' 47" from the pole of the daily axis of rotation.

It must seem, to those persons whose minds are not in a condition of slavery to authorities, a singular proceeding on the part of persons claiming to be astronomers, to endeavour to prove that astronomy is so feeble a science that it fails utterly to explain those changes of climate which are proved to have occurred in the past history of our earth.

Whatever may have been the cause of these changes of climate, say theorists, yet exact astronomy can give no explanation, because M. La Place has proved that the plane of the ecliptic cannot vary more than 1° 21', *therefore* no variation greater than 1° 21' can occur *in the obliquity*.

It is certainly unfair to ignore our earth and its movements in this investigation, but it is a mere repetition of the proceedings of those gentlemen who were taught in the school of Ptolemy, and who attributed wonderful movements to everything in the universe, but denied the movements of the earth.

It is again from the movements of the earth that a solution is to be obtained, not only of the mysterious changes of climate in the past, but also of the supposed changes in the positions of the stars.

CHAPTER XVII.

ANALOGY IN THE SOLAR SYSTEM.

THE course which the earth's axis has traced during the past two thousand years as proved by recorded observations, and due to the second rotation of the earth, is, that at the date 13,544 B.C. the climatic conditions in both hemispheres were such as to produce an arctic climate in winter down to 55° latitude, and an almost tropical climate in summer up to the same latitude. These great alternations of climate annually were simultaneous in both hemispheres of the earth, the northern hemisphere having its winter when the southern hemisphere had its summer, the same as at present.

From the date 21,460 B.C., when the arctic and antarctic circles extended to 29° 30' from the poles, to 13,544 B.C., when these circles reached to 35° 25' 47", the severity of the winters would increase. From this latter date down to about 5000 or 4000 B.C., the severity of the winters would decrease, but the great decrease would occur between about 2000 B.C. and our present time. Thus what has been termed the Glacial epoch of geology, or, more correctly speaking, the *last* glacial epoch, occupied rather more than fifteen thousand years, viz. from about 21,460 B.C. to about 5624 B.C.

During those years the climate of localities greater than

45° or 50° of latitude would not have been so suitable as a residence for the human race as were localities nearer the equator. After the date about 3000 B.C., when the alternations of climate annually became less, then a vast migration of the human race must have occurred from tropical regions through Europe.

But the whole period of one second rotation, acting as it does at present, occupies rather more than 31,600 years. Was there only one great day of this kind, or several days, during which the same conditions prevailed? Does the evidence of geology indicate that only one such great day existed? It appears that the evidence is in favour of a repetition of such days; how many it is for geologists, if they can, to determine.

We now come to a question in which we leave the certainty of geometry, and have to examine probabilities in accordance with the evidence before us.

The daily rotation of the earth takes place round an axis which appears fixed *in* the earth; but this axis changes its direction in consequence of a second rotation, this second rotation occurring at present round two poles, which are 29° 25' 47" from the poles of daily rotation.

We know that when a rigid body rotates rapidly round a fixed axis, but this fixed axis changes its direction, the amount of this change appears to be dependent on the position of the centre of gravity of the rotating body. For example, spin a gyroscope rapidly, and note how the axis of this gyroscope changes its direction, whilst the rapid rotation continues round this same axis. Then attach the smallest object, such, for example, as a pin, to the gyroscope, and the axis of rapid rotation will change its direction in a very different manner from that in which it changed previous to the pin being attached to it. The fact of the pin being attached has slightly altered the position of the centre of gravity of this rotating body, and a different movement of the axis of rapid rotation immediately occurs.

Even a pin, compared to a gyroscope, is a heavy body, but the effects of attaching this pin are *at once* manifested. If we could cause a gyroscope to continue rotating during several days, the head of a pin attached to the rotating sphere would in time produce nearly the same effect that the entire pin produces *at once*.

If we could cause a gyroscope to continue rotating during a year, the most minute change in the position of the centre of gravity of this rotating body would cause a change in the manner in which the axis of rapid rotation alters its direction.

Effects which occur rapidly when a *heavy* body causes the position of the centre of gravity of a rotating sphere to alter its position, occur slowly when a light body causes this centre of gravity to change.

We must now consider the enormous periods of time with which we have to deal in connection with even the second rotation of the earth. This second rotation appears, from the recorded evidence of the past 2000 years, to take place as uniformly as does the daily rotation. The rate of the second rotation is at present $40.9''$ annually; consequently, at this rate, 31,682 years would be required to complete one second rotation.

Taking 366 sidereal rotations of the earth for each year, we should for each second rotation have 11,595,612 sidereal rotations of the earth.

A change, however slight, in the position of the centre of gravity of the earth would produce, during several of these second rotations, a change in the position of the pole of the *second* axis of rotation, and this change might be considerable, although the poles of the axis of *daily* rotation might change their positions in the earth so very slightly as not to be perceived during thousands of years.

In what manner is it possible for the centre of gravity of the earth to change its position, and where is the centre of gravity of the earth at present?

At the present time, at least three-fourths of the land that is above the sea is in the northern hemisphere, and there is at least double as much land above the sea between 15° west longitude and 180° east longitude as there is in the other portion of the northern hemisphere. It is impossible, therefore, under present conditions, that the centre of gravity of the earth can be located in the plane of the equator.

It also follows that if the poles of the axis of daily rotation are equidistant from all parts of the equator, then the axis joining these poles cannot pass through the centre of gravity of the earth, in consequence of this centre of gravity being on a meridian probably 45° east of Greenwich.

It is proved by geology that the relative position of land and water which now exists on earth did not prevail in former times. Continents in the northern hemisphere now above water were in the earlier ages submerged. To submerge continents in the northern hemisphere at the present time would cause the water now in the southern hemisphere to be transferred to the northern hemisphere, and the transfer of this water would cause the centre of gravity of the earth to alter its position.

The alteration of the position of this centre of gravity may have been very slight, but if we dealt with merely three second rotations, we should have more than thirty-four million daily rotations during these three second rotations for an effect to be produced.

Now, a comparatively slight change in the position of the poles of second rotation may produce a very great change in the climatic conditions during a second rotation.

For example, under present conditions the pole of second rotation is 29° 25' 47" from the pole of daily rotation, and hence, during one second rotation, the arctic circle reaches to as far as 35° 25' 47" from the pole, as shown in previous diagrams. If, however, the pole of the second axis of rotation shifted a few degrees owing to a change in the position of the centre of gravity of the earth, we might have the pole of the heavens tracing a circle in the heavens of only 10° or 15° radius, and the pole of the ecliptic might be within this circle, when scarcely any variation of climate would take place during the year for many thousands of years.

Whilst it is almost impossible to cause the axis of a rapidly rotating body to alter its position in this body, it requires but a trifling change in the centre of gravity of this body to cause the axis of rapid rotation *to change its direction in various ways.*

The evidence of geology as regards the last glacial epoch tends to prove that the axis of daily rotation of the earth was situated *in* the earth, where it is at present; but this same evidence proves that an arctic climate extended to about 54° latitude in both hemispheres, and the line of cold ran through America, Europe, and Asia on this parallel, just as is proved by the second rotation of the earth, which reveals the fact that at 13,544 B.C. the arctic circle extended to 35° 25' 47" from the poles, which is to latitude 54° 34' 13".

To conclude, however, that because the poles of the second axis of rotation are now 29° 25' 47" from the poles of daily rotation, they must always have been so, no matter how much or how little the position of the centre of gravity of the earth has varied, is unsound and is impossible.

There may have been several second rotations similar to that now occurring to produce the glacial epoch in its fulness; all that can be stated with certainty is, that the present movement of the earth, carried back some 15,000 years, viz. to 13,544 B.C., would cause the arctic circle to reach to below 54° latitude, not in consequence of any change in the position of the plane of the ecliptic, but due entirely to the movement now taking place in the earth itself.

How small a movement in the position of the pole of the second axis of rotation would produce a very great change in the climate on earth during this second rotation may be understood by aid of the following diagram (Fig. 81).

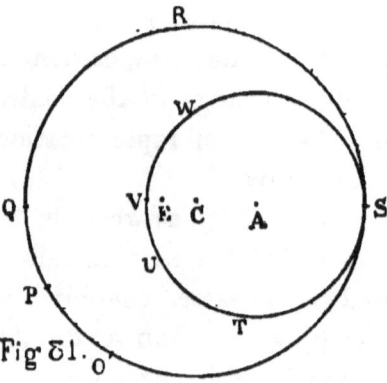

Fig 81.

The large circle represents the course in the heavens which the earth's axis traces at the present time round C, the pole of the second axis of rotation; E, the position of the pole of ecliptic.

During one second rotation, under these conditions, the obliquity varies from S E to Q E, viz. from 35° 25′ 47″ to 23° 25′ 47″. If, however, the pole of the second axis of rotation shifted either rapidly or slowly towards the pole of daily rotation, when this pole of daily rotation was near S, and the pole of second rotation moved from C to A, then the second rotation would carry the pole of daily rotation round the circle S T U V W.

When the pole of daily rotation had reached V, the obliquity would be represented by the small arc V E of only 2° or 3°, and an almost uniform climate would prevail annually over the whole earth during many thousand years.

As the pole of daily rotation was carried round from W to S, the annual alternations of climate would again prevail.

These seem to have been the conditions which prevailed during the formation of coal-beds. When the pole was at V, the uniform climate would allow of even tropical ferns flourishing during thousands of years in high northern regions. When the pole was being carried by the second rotation from W to S, and round to T, the cold of the winters which must then occur would cause these beds of accumulated vegetable matter to be buried by masses of sandstone, shale, and other deposits.

On reaching U, the pole of daily rotation, relative to E, the pole of the ecliptic, would again cause an almost uniform climate to prevail annually over the whole earth, and another accumulation of vegetable matter would take place, to be again followed by the alternations of climate, as the pole of daily rotation was carried round from W to S and T.

The number of beds of coal would indicate the number of second rotations that occurred under these conditions during the Carboniferous period.

The two great conflicts which have occurred in the past history of astronomy, between theorists and the science of geometry, were those relative to the earth being a flat surface or spherical in form, and whether the earth never moved or whether it rotated daily round an axis.

Both the spherical form of the earth and its daily rotation were successfully burked from being acknowledged during many hundred years, due probably to the fact that the merely blind followers of popular theories, and the copyists of the assertions of authorities, have always been far more numerous than sound geometricians and competent reasoners.

Amidst all the controversies in connection with these two problems, it appears singular that the analogy of nature was never examined, and the various celestial bodies of which our earth is a member were never taken as examples.

It may be true that, before the invention of telescopes, all the facts relative to the various bodies in the solar system could not be accurately known, yet enough was known to serve at least as a guide. We, however, now know enough to perceive how powerful this analogy really is.

Every planet and satellite in the solar system is spherical or spheroidal in form: why should the earth be assumed an exception to every other body, and asserted to be a flat disc?

Every planet and also the sun rotate round an axis: why should our earth be an exception, and remain immovable?

From analogy alone the first conclusion of a reasoner would have been that the earth *was* spherical in form and rotated on its axis, not that it was a flat surface and remained stationary.

We have another analogy revealed to us by aid of our telescopes. We know that, whilst the planes of the orbits of the various planets differ only a very few degrees from each other, yet the angle which the axis of daily rotation of each planet makes with the plane of its orbit varies in every conceivable manner.

The axis of Uranus coincides very nearly with the

plane of its orbit. The axis of Saturn is inclined about 62° to the plane of its orbit. The axis of Jupiter is nearly at right angles to its orbit. The axis of Mars is inclined at an angle of about 62° to the plane of its orbit. The axis of the earth is inclined at an angle of about 66° 33' to the plane of its orbit. The axis of Venus is inclined at an angle of about 15° to the plane of its orbit.

This variation in the various angles which the axis of rotation makes with the plane of the planet's orbit is due to some condition in the planet itself; it is not due to the planes of the planet's orbit being inclined to each other at very great angles.

The inclination which the orbit of Venus makes with that of the earth (the ecliptic) is only about 3° 23'; but the inclination of the axis of Venus, as regards its orbit, differs from that which the earth's axis makes with the ecliptic by about 51°. It is thus evident that the position of the planet's axis is the cause which produces these great variations, not any great changes which occur in the planet's course round the sun.

With a good telescope we can perceive on Mars the snows round the pole, which for the time being is undergoing the winter season. As that planet moves round the sun, and the summer season comes to the same pole, the snows decrease, but they increase round the opposite pole, which is then in winter. But the snow-line in Mars extends further from her poles than it does on the earth, because the obliquity on Mars is about 28° instead of 23° 27' 14", as it now is on the earth.

The annual changes of climate on Venus must be excessively severe, the arctic circle extending to within 15° of her equator, and the tropics reaching to within 15° of her poles. A locality on Venus, situated as is England on our earth, would be frozen up in winter, and exposed to intense heat in summer. Here is a cause which would produce such effects as those stated by geologists to have occurred in a modified form during the boulder period of geology on our earth. Yet the orbit of Venus differs from that of the earth by only about 3° 23′.

Referring to Jupiter, we find that scarcely any change of climate occurs annually on that planet, in consequence of the axis of daily rotation being nearly vertical to the plane of the planet's orbit round the sun.

On Jupiter there are not those great annual variations of climate which occur on Venus, or even on the earth. Yet the orbit of Jupiter forms with that of the earth an angle of about 1° 18′ 51″ only.

If geometry were an unknown science, these facts would prove that the angle which the axis of a planet makes with its orbit round the sun is the cause of the varied annual changes of climate which occur on a planet, and any movement in this planet which causes the axis to change its angle with its orbit must cause a variation in the annual changes of climate.

In spite of these facts, theorists assert that the only possible way in which any great changes of climate can occur on a planet is by a change in the position of its orbit round the sun.

We have in the solar system every variety of climate shown by various planets. We have the uniform climate shown annually by Jupiter. We have the alternations of climate shown annually by Mars, which are slightly greater than those which occur annually on earth.

We have those great annual alternations shown by Venus, on which planet the northern and southern hemispheres are, during each winter, subjected to an arctic climate down to 15° latitude, and during summer these same regions are subjected during many weeks to a vertical sun at midday, and the sun at an altitude of several degrees at midnight.

There are no facts revealed by geology, as regards climate on earth in past ages, that do not find a ready explanation by such conditions as now prevail between the state of Venus and that of Jupiter.

Are we to assume that the axis of Jupiter has no movement? or that it traces a circle round a point about 2° from its pole, by means of "a conical movement of its axis"? Are we to assert as a dogma that the axis of Venus traces a conical movement round a point 75° from the pole of daily rotation of Venus? or are we to take a wider view of this problem, and state that the assertion of the modern theorist, who claims that no change in the obliquity or extent of the arctic circle can possibly occur, except by means of a change in *the orbit* of a planet, is a statement untrue as regards geometry, and is contradicted by facts known in connection with other planets in the solar system?

If the axis of Jupiter were caused to change its direction by means of a second rotation some 10° or 15°, we should have conditions on that planet similar to those which prevailed on earth during the formation of our coal-beds.

The actual conditions now prevailing on Venus are such as to produce a glacial period far more severe, and extending to lower latitudes, than that which geological evidence proves existed on earth: and why? Because the arctic circle in Venus extends to 75° from her poles; whereas, during the last glacial period on earth, the arctic circle extended to only 35° 25' 47" from the poles, as proved by the course which the axis traces and has traced on the sphere of the heavens during the past two thousand years.

In the planet Uranus the axis is nearly coincident with the plane of its orbit, not because the plane of this planet's orbit is nearly at right angles to that of Jupiter, but because, from some condition in Uranus (probably the position of the centre of gravity), the axis of daily rotation is at present inclined at a very small angle to the plane of the planet's orbit.

The plane of the orbit of Uranus differs from that of Jupiter less than 1°; yet in Jupiter the arctic circles do not extend 3° from the poles, whilst in Uranus they extend more than 80°—a result in each case due *to the position of the axis of daily rotation relative to the plane of the planet's orbit round the sun.*

With such facts before us, we may with benefit reflect on the assertions of the modern theorists, who tell us that no possible change in the obliquity or extent of the arctic circle can occur, except in consequence of a change in the position of the plane of the earth's orbit round the sun. In spite of this positive statement, it really seems that the planets Venus, Jupiter, and Uranus have found out a method of accomplishing this impossibility.

It is also a fact that, if the pole of the heavens trace *a circle* round any point in the heavens other than the pole of the ecliptic *as a centre*, the earth itself can accomplish the same impossibility.

The satellites of the planet Uranus offer an amusing example of how a very simple geometrical law will sometimes effectually puzzle those persons whose minds have lost clearness of perception in consequence of being overcrammed with dogmatic theories, or are in such a state as to render them capable only of following certain authorities.

It has been stated that the axis of the planet Uranus is nearly coincident with the plane of its orbit. The satellites of Uranus revolve round their primary planet in circles which are inclined but slightly to the equator of Uranus; but the orbits of these satellites are inclined at

ANALOGY IN THE SOLAR SYSTEM. 267

an angle of about 78° 58' to the plane of the earth's orbit. The relative positions of Uranus and the earth (not drawn to scale for want of space) will be understood from the following diagram (Fig. 82).

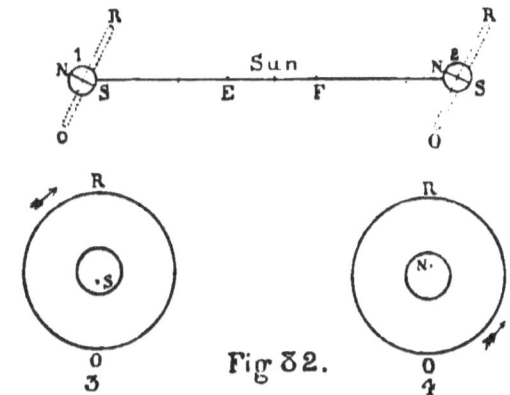

Fig 82.

No. 1 represents the position of Uranus, the axis of daily rotation being shown by N S; N the north, S the south pole; E, the position of the earth. The satellites of Uranus revolve round the planet from O to R.

From the earth at E these satellites will appear to travel round the planet as shown in No. 3 by the arrow, that is, they travel round Uranus in the same manner as one deals at whist—from left to right.

When Uranus is at No. 1 and the earth at E, the orbit of these satellites, if projected on the plane of the ecliptic, might be stated to move from east to west, because Uranus itself would be south, and seen best at midnight, being then in opposition to the sun.

The south pole of Uranus would, under the above conditions, be turned towards the sun and earth, and the same effects would be observed as regards these satellites as can be noticed by an observer in the southern hemisphere of the earth, who would perceive the moon, night after night, move from left to right about 13°.

When Uranus had been carried over 180° of his orbit,

and had reached No. 2, then the north pole of Uranus would be turned towards the sun and the earth at F, and the satellites, still revolving as before, would now appear to travel round Uranus as shown in No. 4, viz. in the opposite way in which one deals at whist, and apparently in the opposite manner in which they before revolved. If the course of these satellites were projected on the plane of the ecliptic, when the north pole of Uranus was turned towards the earth, they would then be said to move from west to east, not from east to west, as they appeared to move when at No. 1.

It is a simple geometrical law, that if the axis of a planet nearly coincide with the plane of the ecliptic, this planet will sometimes present its north, sometimes its south, pole to the earth. If the satellites of this planet revolve round, or nearly round, the equator of this planet, they will, when the south pole is directed towards the earth, appear to move round the planet in the same manner as one deals at whist.

When the north pole is directed towards the earth, they will appear to move round their primary in the opposite direction to that in which one deals at whist.

There is nothing wonderful in these changes; it scarcely deserves the name even of a geometrical problem—it is a self-evident fact.

Now let us examine what has been made out of this self-evident fact by theorists.

Uranus was discovered on the 13th of March, 1781, by Sir William Herschel, by aid of a large telescope. He announced, also, that this planet possessed satellites, which presented a remarkable contradiction to everything else in the solar system.

In Sir John Herschel's "Outlines of Astronomy," Article 552, this wonderful "contradiction" is stated as follows :—

"The orbits of these satellites offer remarkable and, indeed, quite unexpected and unexampled peculiarities. Contrary to the unbroken analogy of the whole planetary system, whether of primaries or secondaries, the planes of their orbits are nearly perpendicular to the ecliptic, being inclined no less than 78° 58' to that plane, and in these orbits their motions *are retrograde;* that is to say, their positions when projected on the plane of the ecliptic, instead of advancing from *west to east,* round the centre of their primary, as is the case with every other planet and satellite, move in the opposite direction."

When the south pole is directed towards the earth, of course the satellites will appear to move in this manner, just as the moon in the southern hemisphere appears to travel each night from left to right; but this left to right in the southern hemisphere is not from east to west, it is from west to east.

A trifling oversight was committed, viz. that of not noticing that the south pole of Uranus was turned towards the earth in 1781, when it was discovered.

Here, however, was a grand opportunity for the theorist. The satellites of Uranus (so it was asserted) move differently from everything else in the solar system. This erroneous statement was copied and recopied in every book on astronomy, and came to be accepted as an absolute fact, upon which theories could be firmly built.

The pope of theorists examined the supposed fact, and after due deliberation announced that, before the whole solar system, consisting of a sun and various planets and satellites, was gradually shrivelling up from the cloudy and nebulous condition in which it had first existed, an interloper from some other nebulous system, which was twisting round in the opposite direction, must have forced itself among the orderly planets belonging to the sun.

Thus Uranus and his satellites were looked upon as an example of some of those American waltzers who practise "the reverse," and who waltz round in the opposite direction to that in which all the other waltzers rotate, and thus cause a considerable amount of confusion in the usually orderly proceedings of the majority of waltzers.

That a planet whose axis was situated very nearly in the plane of its orbit, should present its south pole to the earth when in one part of this orbit, and its north pole, when 180° from the first-named position, should be directed towards the earth, was too simple a fact to be comprehended.

Theorist after theorist invented some wonderful explanation in order to account for that which did not occur. Copyist after copyist repeated in their books of astronomy the assertion that the satellites of Uranus moved round their primary planet in a different manner to that in which every other planet and satellite moved.

Strange to say, the same error is even now copied by gentlemen, who, by the aid of a pair of scissors, produce new books on astronomy, although nearly thirty years have elapsed since I first called attention to the fact that the assertion relative to the satellites of Uranus was a mare's nest.

It was about thirty years ago that, having occasion to examine this problem relative to the satellites of Uranus, I at once perceived that Uranus at one point in its orbit would present its south pole to the earth, and about forty-two years afterwards would present its north pole to the earth. When the south pole was presented to the earth the satellites would move round Uranus in the same manner as one deals at whist. When the north pole of Uranus is seen from the earth, then the satellites move round in the opposite direction to that in which one deals at whist.

Surprised that so very simple a geometrical law should have puzzled the learned gentlemen who had written on this subject, I prepared a paper with diagrams, in order to explain the laws which produced the effects.

Being young and inexperienced at that time, I believed that the solution of this supposed mystery would very likely be well received by a society formed for the ostensible purpose of eliciting truth and advancing real science.

I was not prepared for the results which followed the reading of my paper, and the proofs I gave of the true cause of the apparent movements of the satellites of Uranus.

Instead of the diagrams proving the truth of my conclusions being discussed, or even examined, it was stated that it was a piece of unwarrantable impertinence for a mere artillery officer to question the infallibility of those authorities, who had stated what they had about these satellites. If the explanations which I had given were correct, of course the great authorities would themselves have found this out; and I ought to be ashamed to show my audacity by presuming to think I had found a solution to that which had been a mystery to the greatest theorists.

A senior officer, who was more successful in holding on to the skirts of those who possessed patronage than he had been for any exhibition of a knowledge of science, considered it his duty to write me an official letter, in which he stated that he should report me to superior authority for having so far forgotten myself as to question the accuracy of the admitted authorities of the time in matters of astronomy, and that he should recommend that I be removed from my appointment.

In the year 1862 (some years after my first paper) I wrote a popular work on astronomy, termed "Common Sights in the Heavens." In this work I repeated my proof about the satellites of Uranus.

In one or two of the modern works on astronomy it is now stated the peculiar movements once supposed to occur in connection with the satellites of Uranus *may* be due to the position which the axis of Uranus occupies relative to its orbit.

In a few years more it will probably be stated in books that the satellites of Uranus move from west to east round their primary in exactly the same manner as do all other satellites in the system.

This proceeding is one among many, and is a specimen of that which is done by a certain class of individuals, who pose before the public as earnest workers for the cause of truth and real science, but who have a firm conviction that it pays better to agree with the leading authorities of the time, than to really fairly examine the accuracy of any novelty, which may be opposed to the views of those officials who have the power of dispensing patronage.

The reader may now perceive that analogy alone might have shown that, as all the planets in our system were spherical, it was not such a very outrageous idea to conclude that the earth also was spherical. Yet the theorists in olden times asserted that the earth was flat, and those who claimed that it was spherical were persons who were to be ridiculed. The total absence of any knowledge of geometry, shown by the believers in a flat earth, was considered of no consequence as affecting their mere opinions.

The fact that the sun and all the planets rotate round an axis would, from analogy alone, one would think, have caused reasoners to suspect that the earth also rotated. Such, however, was an idea which theorists at once opposed. If the earth rotated, all the water in the oceans would be flung off just as mud is flung off from a rapidly rotating wheel; the water of the ocean is not flung off, *therefore* the earth cannot rotate;—was the argument of the theorist.

We have now a third analogy. The planes of the orbits of the various planets in the solar system differ but slightly from each other. Yet the angle which the axis of daily rotation of these planets forms with the orbits differs from nearly 90° to about 2°.

Every variety of climate, therefore, exists on various planets belonging to the solar system. In Jupiter the climate is uniform for each locality during the whole year. In Venus the alternations annually are from an arctic to a tropical climate for all latitudes from 15° to the poles of Venus. These climates are due to the angle which the axis of rotation of these planets forms with the plane of the planet's orbit round the sun.

The facts revealed by geology prove that in the past history of the earth every variety of climate has prevailed. At one time the climate in arctic regions was suitable to the growth of tropical plants, the remains of which are now found embedded in the soil, proving that they grew where they are now found. The conditions now existing in Jupiter are suitable to such results.

We have from geology evidence of an arctic climate having existed on earth down to 54° latitude in each hemisphere during the glacial period. In Venus we have exactly those conditions at present existing in an intensified form.

With such facts before us, what could be more natural or simple than to believe, from analogy alone, in the possibility of the *direction* of a planet's axis of daily rotation changing considerably, and thus causing an extension or contraction of the arctic and antarctic circles and of the tropics?

If the evidence of the past two thousand years proved that no change whatever had taken place in the direction in which the earth's axis pointed, it would not follow that,

T

during the endless ages of the past, some conditions which do not now exist might have caused this axis to change its direction.

What, however, are the facts? It is known that during the past two thousand years the earth's axis has uniformly changed its direction, and at the rate of about 20·09" annually. It is known that during all these years the pole of the heavens (which is that point in the heavens towards which the axis points) has decreased its distance from the pole of the ecliptic, causing a corresponding decrease in the extent of the arctic circle and tropics.

It was, therefore, the duty of men of real science to investigate what this movement of the earth really was which caused the axis to change its direction, and also to examine the true course of the earth's axis on the sphere of the heavens. More especially was this a duty when the position of the axes of other planets indicated every variety, and when the evidence of geology proved that every variety of climate had in past ages prevailed on earth.

What did theorists do? They ignored the movements of the earth itself, and set to work to invent theories as to how much or how little *the orbit* of the earth might vary. Such a proceeding was as inapplicable to the problem under inquiry as though these persons had set to work to find out how much the orbit of Jupiter or that of Venus could vary, in order to discover the angles which the axes of these planets made with the planes of their orbits.

Having by their theories imagined that the plane of the earth's orbit could vary only 1° 21' from a mean position, they actually presumed to assert that this was the limit of variation in the extent of the arctic circle on earth.

Some years afterwards, however, this theory, which had been bowed to as reverentially as though it had emanated

ANALOGY IN THE SOLAR SYSTEM.

from the Vatican, is found to be wrong; and another theory is invented, and again bowed to, viz. that the plane of the ecliptic can vary, not 1° 21' only, but as much as 4° 52', and *therefore* this 4° 52' is now imagined to be the limit of variation in the arctic circle, *because* the plane of the ecliptic is supposed not to be able to vary more than that amount.

A chorus of all the learned theorists formerly sang 1° 21' only; now they sing 4° 52' is all the change that can possibly occur in the extent of the arctic circle, *because* the plane of the ecliptic cannot vary one second more than 4° 52'. These assertions are made whilst Venus and Jupiter are giving examples of an entirely contradictory character.

But the most marvellous fact connected with these theories is, that not only is the movement of the earth in connection with this problem utterly ignored, but the actual movement of the earth is not considered of sufficient consequence to receive an accurate description.

Theorists assert that the earth's *axis* traces a conical movement during 24,868 years. Which pole of the earth remains fixed whilst this conical movement is being made? This item has not been considered of sufficient importance to be even mentioned.

How is the zenith of 51° north latitude and of 61° north latitude affected by this conical movement, as regards six hours, twelve hours, and eighteen hours right ascension?

Of course, all such details are assumed to be known, for has not a learned critic, in a periodical devoted to scientific subjects, assured his readers that I am evidently not acquainted with the important fact that "the joint action of the sun and moon on the earth's protuberant equator makes 'a shift' in the earth's axis"?

The action of the sun and moon *may* make a pair of

boots in the earth's axis, but it would be a matter of more practical value to discover and define what is the actual movement of the earth which *causes* the change in direction of the axis.

This actual movement is a second rotation round an axis, the poles of which are now situated in the heavens 29° 25′ 47″ from the poles of daily rotation; and it is not a conical movement of the whole axis round the pole of the ecliptic as a centre, as has been erroneously asserted by theorists during the past two hundred years.

If any competent geometrician, possessing sufficient moral courage to undertake an original investigation, had been guided by the analogy in the solar system, and by the facts in geology, to examine what were the possible and what the probable movements of *the earth* which occurred in connection with the change in direction of the axis, he would have discovered that a second rotation must take place, in order to account for the observed facts.

This movement might have been discovered a hundred years or more in the past, with the result that calculations could have been made with ease or accuracy, which even now can be arrived at only by perpetual observation from year to year, at the cost of many thousands of pounds per annum.

What, however, was the course adopted by the so-called scientific authorities of the past? Apparently contented with the superficial assertion that the earth's axis had a conical movement, they never seemed to think it of any importance to state which pole remained fixed whilst this cone was traced. The zenith of the terrestrial pole moved, as found by endless observation, about 20·09″ annually towards the first point of Aries. The knowledge of this one fact seemed to be considered quite sufficient, and consequently no trouble was taken even to inquire how the zeniths of various localities on earth were displaced annually by the same movement of the earth which causes the zenith of the pole to change its position 20·09" annually.

The most amusing proceeding of all, however, is that, whilst this ignorance of the detail movements of the earth existed, theorists by the score set to work to invent theories to account for that which did not occur. French, German, and English theorists invented the most wonderful problems to explain why the earth's axis always traced a cone round the pole of the ecliptic as a centre, not one of these theories ever hinting as to which pole of the earth remained fixed whilst the cone was traced. How it was possible that the earth's axis traced *a circle* round *a movable centre* was not even considered worth mentioning. All these theories, however, were based on the same assumption, viz. that the joint action of the sun and moon on the earth's equator caused "*a* conical movement of the earth's axis."

This is the theory which certain scientific copyists continue repeating in the present day, and which they seem to imagine is amply sufficient to explain all those movements of the earth with which they are unacquainted, and of which they prove they are unacquainted by claiming that repeated observations are still necessary, in order to frame a catalogue of stars for a few years in advance.

When the theorist can, by a geometrical calculation based on one observation made of a star, calculate the mean polar distance of this star for fifty or a hundred years, past or future, he may have established some claims to be an authority whose mere opinion ought to be treated with deference.

When, however, no such calculations are known or comprehended by the theorist, and when we find that theories are invented to account for that which does not occur, and that a vast mystery is assumed to take place because a planet sometimes presents its south, at other times its north, pole to the earth, then the mere opinion of such gentlemen cannot be looked upon by a reasoner as infallible. And although the second rotation of the earth may not be agreed to—and naturally it will not be agreed to—by the present theoretical authorities, yet, when calculations can be made by aid of a knowledge of this second rotation which are now quite beyond the knowledge of those gentlemen who claim to be authorities, it remains for the reader to judge whether it is of much importance to the cause of truth, whether such authorities do or do not agree in their opinions as regards the fact of the second rotation of the earth.

In the past history of astronomy, the opinions of authorities at various dates were quite opposed to the earth being spherical, and they also denied the possibility of its daily rotation. Mere opinions, however, can but retard truth during a few years; facts and reason sooner or later uproot prejudice and even vested interests. When the mere opinions of certain authorities will enable them to calculate the polar distance of a star for fifty years from one observation only, then they will be able to give a practical proof that these opinions are of some value. Such, however, is not the case at present.

If geometricians and reasoners would free their minds from the dogmatic theories which now prevent facts from being perceived, they would find open to them a vast field of untrodden ground which they could investigate, and on which they could make as great an advance in the science of astronomy as was made by those reasoners who accepted the daily rotation of the earth, and freed their minds from the errors of the system of Ptolemy.

CHAPTER XVIII.

OBJECTIONS OF THEORISTS.

THOSE persons who have studied the past history of astronomical progress can scarcely fail to come to the conclusion that, in past ages at least, the human mind was in such a condition of slavery to routine and authority, that it was incapable of forming a correct opinion on any novelty that was presented to it. A novelty may be true, or it may be false, but in both cases the objections urged against it ought not to be more absurd than the novelty itself.

When the earth was said to be spherical in form, one of the objections was that all the water of the ocean would run off if it were round, because water would not stay on a small ball. Another objection was that people on the opposite side of the earth would be walking with their heads downwards; and as no man could walk on the ceiling of a room with his head downwards, *therefore* the earth could not be round.

In the absence of a knowledge of the laws of gravitation, neither of these objections could be claimed as utterly absurd, and it speaks well for the reasoning faculties of the ancients that, in spite of these and similar arguments, they agreed that the earth was spherical, and was not a flat surface.

In the present day, when we have the electric telegraph and reliable clocks, we have such proofs of the earth's spherical form that we are independent even of geometrical proofs.

A person in India, say on March 21, can telegraph instantly to a person in England at the instant that the sun rises. This telegram will reach England at midnight. The person in India can telegraph to a person in England when the sun is south in India, and this telegram reaches England just as the sun is rising. A person in England can telegraph to a person in Canada, in 90° west longitude, when it is midday in England; this message reaches Canada at daybreak. How can the flat-earth theory explain these facts? It avoids them by denying them, or by asserting that the sun is only a few miles from the earth, and that the variable distance of the sun explains the facts. Here, again, the flat-earth theory is untenable, because if the distance of the sun were so very slight, it would appear scarcely half the size at sunrise that it does when on the meridian, as it must be much nearer when on the meridian than it was at sunrise.

The evidence from geometry as regards the moon is, however, undeniable; but the ancients, like some of the modern theorists, did not appear to be acquainted with geometry, or at least, if acquainted with this science, they did not seem capable of applying it.

When the daily rotation of the earth was again brought into notice by Copernicus, and afterwards by Galileo, the authorities of the day again opposed this innovation, and used arguments against it. Many of these arguments were at the time, no doubt, considered powerful and unanswerable. Some were absolutely ridiculous, and showed how feeble was the intellectual calibre of those who urged them.

In more modern times we find that the arguments used by objectors to novelties are far more ridiculous than any that the ancients used—a result probably due to the fact that some of the moderns were blinded intellectually by dogmatic theories in which they had unbounded faith, and any facts which interfered with these theories must be ignored in every possible manner.

Thus, when by aid of a telescope Galileo had discovered the satellites of Jupiter, he was attacked by two learned astronomical authorities of the time, viz. Libri of Pisa, and a gentleman named Sizzi.

The latter considered it his duty to give an harangue against Galileo, in which he made the following assertions: "There are seven windows given to animals in the domicile of the head, through which the air is admitted to the tabernacle of the body—viz. two nostrils, two eyes, two ears and one mouth. So in the heavens, as in a microcosm, or great world, there are two favourable stars, Jupiter and Venus; two unpropitious, Mars and Saturn; two luminaries, the sun and moon; and Mercury alone undecided and indifferent. From these, and from many other phenomena of nature, which it were tedious to enumerate, we gather that the number of the planets is necessarily seven. Moreover, the satellites are invisible to the naked eye, and therefore can exercise no influence over the earth, and would, of course, be useless; and therefore do not exist."

This argument was no doubt at the time considered unanswerable, and to exhibit the profound knowledge of Sizzi, and the ignorance which Galileo displayed, in not being aware that because there were only seven holes in the head, there could not be more than seven planets and satellites in the heavens.

Another illustrious astronomical authority, viz. Huyghens, talked in a similar manner. He, having discovered one satellite of Saturn, announced that the solar system was then complete, as it consisted of six planets and six moons; and consequently that no more planets or satellites would ever be discovered.

Some of the philosophers refused to look through so diabolical an engine as a telescope. This instrument, having proved that their preconceived theories and opinions were wrong, hurt their self-esteem as authorities and teachers of what was imagined to be "exact astronomy."

The exhibition of this frame of mind, viz. the firm belief in theories and the refusal to even look through a telescope, is very instructive. In scientific investigations it is necessary that we should be acquainted to some extent with the peculiar habits of certain types of minds, especially when these minds are very firm in their opinions, but somewhat feeble in their reasoning, and are very deficient in facts.

In the case of the satellites of Uranus already mentioned, it is evident that there was a readiness to accept without investigation the statement of an authority. There was something wonderful and mysterious said to occur in connection with these satellites; they were said to move from east to west, instead of from west to east, like other satellites in the solar system. The love of wonder possessed by certain individuals caused this assertion to be at once accepted as true, and it was promulgated far and wide as a reality.

The simple geometry of the question was never looked at. That Uranus must present the south pole to the earth when in one part of its orbit, and its north pole when in the opposite part of its orbit, with the apparent results already mentioned, was entirely overlooked. Consequently, this supposed mystery was copied and recopied by astronomical authorities during more than eighty years.

In referring to the remarks which have been made by certain individuals relative to the second rotation of the earth, as explained in this and my other works on the subject, it is scarcely necessary to refer to those stereotyped objections which seem to have been urged during all ages against each discovery.

If this were true, it would have been discovered long ago.

Is it likely that all the authorities are wrong and one person right?

The authorities do not agree with this idea of a second rotation.

These, and many similar opinions have been expressed. The history of the past is almost a sufficient reply to such remarks.

The first objection applies equally to every discovery that has been made.

It *is* likely, and must be a fact when a discovery occurs, that the authorities who were unacquainted with this discovery were wrong. The authorities were all wrong when they affirmed that the earth could not move, and that, because animals had seven holes in their heads, therefore Jupiter could not possess satellites.

The authorities who refused to look through a telescope did not agree with the statement that Jupiter possessed satellites, and those persons who have not, or will not examine the facts which are proved by the second rotation of the earth occupy a similar position.

Every individual who has examined, without prejudice or preconceived opinions, the models and calculations proving this second rotation, has stated that it is convincing; and many persons, whose preconceived opinions were opposed to the idea, have, when they have examined the facts, been satisfied of its accuracy.

Such objections need not be further replied to, because the second rotation of the earth is a problem which can be demonstrated and proved; and calculations can be made by a knowledge of this second rotation which have hitherto never been supposed possible. If it be considered that mere opinions are to successfully defeat facts, then there is an end to all scientific progress, and had such been the case formerly, Jupiter's satellites would have long been considered optical delusions, produced by that diabolical machine, the telescope.

When a novelty is brought forward in connection with any science which has been imagined to have arrived at a state of perfection, it appears as though this novelty completely disarranged the mental machinery of those persons who were the supposed authorities at the time. This disarrangement is then exhibited in various ways but more particularly in bringing to bear on the novelty arguments which are so palpably erroneous, that any reasoner, even if unacquainted with the science of the subject under discussion, would without hesitation pronounce as childish.

The reader who has made himself acquainted with the facts presented to him in the preceding pages, will realize how very slight is the difference between the theory which has been believed in during the past three hundred years, and the movement of the earth, which proves to be a second rotation.

The theory hitherto believed in was that the earth's *axis* traced a cone, and consequently the pole of the heavens traced a circle in the heavens, round the pole of the ecliptic as a centre. If this movement really took place, it must follow that either the south or north pole of the earth remained fixed, whilst that pole of the earth which was not fixed as regards this movement described the base of the cone.

Which pole, however, remained fixed, and which described

the cone, was apparently considered too trifling a fact for theorists to trouble themselves about.

A knowledge of the second rotation of the earth teaches that neither pole remains fixed as regards this movement; it is the centre of gravity of the earth which remains fixed, whilst the two half-axes of the earth describe cones.

Thus the two half-axes during one second rotation perform similar movements to those which a line passing through the earth's centre performs during each daily rotation. A line from any point on the earth's surface to the earth's centre traces a cone every twenty-four hours, the zenith of each locality tracing a circle in the heavens during the same time. When a sphere rotates round an axis, each point on this sphere must trace a circle, except those two points which are the poles of the axis of rotation.

This is a mere elementary geometrical law, with which one would imagine every tyro even in science must be acquainted.

It has been stated, during the past three hundred years at least, that the earth's axis traces a circle in the heavens, but that it traces this circle round a point *as a centre*, from which it has decreased its distance during two thousand years at least.

When the movement of the earth herein demonstrated was, some eighteen years ago, first brought into notice, the statement that there might be something with which certain authorities were not acquainted, was regarded as a piece of audacity which must be at once put down. One of the arguments intended to disprove this movement was the following, which, being undoubtedly the most brilliant intellectual effort of which the propounders were capable, ought to be treated with due respect.

It was stated that it was a well-known fact in astronomy that all the planets revolved round the sun in elliptical orbits; none of them revolved round the sun in circles. This, stated the learned objectors, is a well-known law. Consequently (they continued), how is it possible that the earth's axis can trace *a circle* in the heavens? It would be opposed to all the laws of gravitation if the earth's axis traced a circle in the heavens. Therefore we reject as impossible this new theory, and adhere to that which has hitherto been accepted as satisfactory, viz. *that the earth's axis traces a circle in the heavens*, round the pole of the ecliptic as a centre!

This remarkable effort of intelligence was considered so profound, that the objection was copied and recopied in various articles, the writers of which used their best endeavours to prove that the second rotation could not occur.

In the *Quarterly Journal of Science*, the late Mr. Belt referred to this objection as a very serious one, but which might be overcome if the earth's axis traced an ellipse, and not a circle.

After duly considering this objection, it is very difficult to accept it as quite as crushing as its inventors seemed to imagine. The exact analogy between the forces which act on a planet revolving round the sun, and the geometrical laws which must occur in connection with the rotation and second rotation of a sphere, do not by any means appear so closely allied as those which it was imagined existed between seven planets in the solar system and seven holes in an animal's head.

Geometry is a science which certain theorists assert can prove nothing; yet call this science by some other name, and it is a fact that the zenith of a locality on earth traces every twenty-four hours a circle in the heavens, round the pole of daily rotation as a centre, in spite of the planets in their orbits round the sun tracing ellipses.

A second rotation of the earth must also cause the

OBJECTIONS OF THEORISTS. 287

zenith of the pole of daily rotation (termed the pole of the heavens) to trace a circle in the heavens, round the pole of the axis of second rotation as a centre; for the same reasons that the zenith of a locality traces a circle on the sphere of the heavens each twenty-four hours owing to the daily rotation.

Certain learned gentlemen have asserted that the pole of the heavens cannot, owing to a second rotation of the earth, trace a circle in the heavens round a point 29° 25' 47" from it, because the planets all move in ellipses round the sun, and that therefore *a circle* cannot be traced by the pole of the heavens; it must be an ellipse. And to prove how sound is their objection, they then state that *therefore* they must adhere to the present accepted theory, viz. that the earth's axis traces *a circle* in the heavens round a point as a centre, from which point the circumference continually decreases its distance.

These statements, although put forward by well-known authorities, are of so very contradictory a character, that a mere reasoner feels disposed to ask whether these gentlemen really knew what they were writing about, or whether the fact of their having been supposed during many years infallible authorities had not caused them to slightly neglect those laws of reason and geometry which must be applied to the exact sciences.

It is a well-known fact in astronomy, say another class of objectors, that the joint action of the sun and moon on the protuberant equator of the earth causes a shift of the earth's axis, and *therefore* this idea of the second rotation of the earth is ridiculous. Why?

Is not this similar to putting the cart before the horse? A theory is invented to account for something, but it has never been yet stated what this "something" really is.

The earth's axis traces a cone, say theorists, but whether

the south or north pole of the earth remains fixed in order that this cone be traced, has never been mentioned. How the zeniths of various localities are annually affected by this movement has never been mentioned. Whether the movement of the two poles over arcs of 20·09" annually are the only changes, and whether the zenith of a locality in latitude 60° 34' 13" north is similarly affected, as regards meridians of six hours and eighteen hours right ascension, have not been subjects which have troubled the minds of theorists, who were fully satisfied when they accepted the statement that the joint action of the sun and moon did "something."

In order to make accurate calculations, we must find out what this something really is; we must be able to give the exact course on the sphere of the heavens which the earth's axis traces; we must define how each zenith is affected by that movement which causes the earth's axis to change its direction; we must be able to state that certain portions of some meridians are retarded (when we refer to the daily rotation) by this movement, whilst other parts of the same meridian are accelerated. In fact, we must define what the exact movements of the earth really are which accompany, or, as is really the case, *cause* the change in direction of the axis. To do this is of far more practical importance than to be able to repeat like parrots certain theories which it is assumed cause a vague something.

If the true movement of the earth be known, the most simple calculations can be made as regards the future position of the stars; this fact is proved in these pages. At present the true movements of the earth are not known to theorists, consequently some ten thousand pounds per year are required for one observatory alone, for the incomes of a staff employed in perpetually observing stars, etc., in order to frame a catalogue for two or three years in advance.

One can scarcely accept, as a powerful objection against the second rotation of the earth, the sing-song statement that the joint action of the sun and moon on the protuberant equator makes "a shift" of the earth's axis. A reasoner naturally desires to investigate what the actual movements of the earth really are, instead of being contented with vague theories assumed to account for still more vague movements.

The whole difference between that which occurs in connection with the second rotation of the earth, and "something" which has hitherto been assumed to occur, is mainly that the second rotation gives all the details of this movement, whilst hitherto theorists have given no details whatever, but have been contented with stating that the earth's axis changes its directions, and that the axis traces a cone during about 25,000 years.

One of the most remarkable facts in connection with the opposition offered by certain individuals to the investigation that I have undertaken as regards the detail movements of the earth is the misapprehension which these persons seem to possess as to what the problem really is. There appears to be an incapacity to comprehend that it is a practical and useful proceeding to examine in detail what the movement of the earth really is which accompanies the change in direction of the earth's axis. From the written remarks of certain gentlemen, it appears that they imagine that this geometrical investigation is some vague theory, whilst to leave this movement of the earth undefined and deficient in every detail is sound and exact science! The problem is a very simple one, and it is not much to ask of those learned theorists who claim to tell us the cause of everything, if we request them to show, by the aid of a small globe, how the various parts of the earth have moved during the past 180 years in consequence of the pole of the heavens having changed its direction 1° during that period.

The zenith of the terrestrial pole has changed its position 1°, the pole having traced an arc 1° in length on the sphere of the heavens, and nearly towards the first point of Aries.

The pole itself is uninfluenced by the daily rotation, and the course which this pole has traced is clearly perceived, but the zenith of all other localities on earth is affected by the daily rotation, and any movement produced on each zenith is mixed up with and partially concealed by the daily rotation. We must therefore separate these two movements, and deal only with that which occurs in connection with the change in the zenith of the pole, and then examine how this movement affects the daily rotation.

Here is a globe representing the earth (Fig. 83). N the north, S the south, pole; A x B, a parallel of latitude of 70°; D C, a parallel of latitude of 51°; E, a parallel of latitude of 20°; F, the equator; and N x D S, a meridian of twelve hours right ascension.

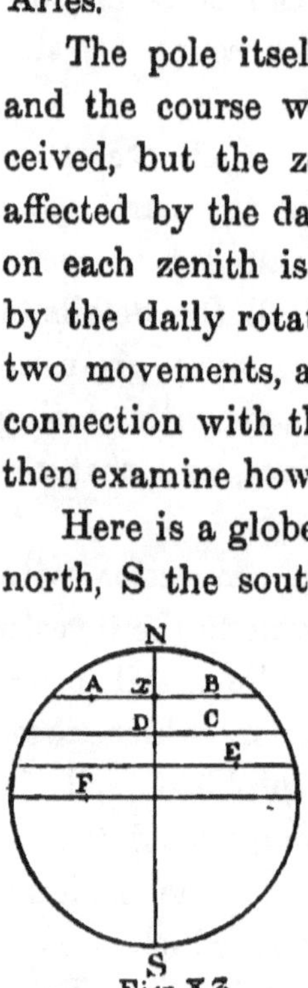

Fig 83.

The pole N is carried over an arc of 1° during 180 years, directly *down* (as we may term it) the meridian on the opposite side of the sphere, by some movement of the earth hitherto assumed to be accurately defined by calling it "a conical movement of the earth's axis." Will some or any learned theorist mark on this diagram where the points A, x, B, D, C, E, and F, will be carried by that same movement, which has caused the pole to be carried 1° down the meridian on the opposite side of the sphere? What is the length and what the direction of the various arcs over which these points are carried, is a fair question for inquiry.

It may be an interesting question for the student or the reader to put to those gentlemen who claim to tell us what it is impossible that the earth under certain assumed conditions can do, but who have hitherto failed to state what the earth does do and has done as regards its movements during the past two thousand years.

Considering that the detail movements of the earth, which accompany the change in the zeniths of the poles of 20·09″ annually, have never been defined or even hinted at by theorists, it is scarcely likely to add to the scientific reputation of certain gentlemen to find them treating this important problem in such a jaunty, offhand manner as they too frequently have treated it.

In a work lately published entitled "Discussions on Climate and Cosmology," the author, Dr. James Croll, devotes seventeen lines to the subject of that movement of the earth to which reference has been made in this and in my former works.

In the first two sentences of the paragraph, Dr. Croll writes as follows: "The theory of a change in the obliquity of the ecliptic has been appealed to. This theory for a time met with a favourable reception, but, as might have been expected, it was soon abandoned."

I am at a loss to understand what information this writer wishes to convey to his readers by these two sentences.

If it were stated that some three hundred years ago Copernicus, and afterwards Galileo, suggested that the rising and setting of the various celestial bodies would be explained by a daily rotation of the earth, but that "this theory, as might have been expected, was soon abandoned," we should not be disposed to attribute any great intellectual capacity to those learned gentlemen who so soon abandoned this theory.

Does Dr. Croll mean it as complimentary, or the reverse,

that the variation in the obliquity was so soon abandoned, whilst at the same time the actual movements of the earth had never been defined?

Any person who has carefully examined the details of the second rotation of the earth, and has proved that most important and accurate calculations can be made by aid of a knowledge of this movement, also that hitherto no such calculations have been possible, and no details of the movements of the earth have ever been given, will attribute the "soon abandoned" to the same cause as that which led to the daily rotation of the earth being abandoned. It is, however, doubtful whether this is the exact meaning which Dr. Croll intended to convey to his readers.

The remainder of the paragraph which this writer devotes to the variation in the extent of the arctic circle is very interesting, he states.

"The researches of Mr. Stockwell of America, and of Professor George Darwin and others in this country, have put it beyond doubt that no probable amount of geographical revolution could ever have altered the obliquity to any sensible extent beyond its present narrow limits. It has been demonstrated, for example, by Professor George Darwin that, supposing the whole equatorial regions up to latitude 45° north and south were sea, and the water to the depth of two thousand feet were placed on the polar regions in the form of ice—and this is the most favourable redistribution of weight possible for producing a change of obliquity—it would not shift the arctic circle by so much as an inch" "Climate and Cosmology," p. 4.

The obliquity of the ecliptic, found by repeated observations at the early part of the Christian era, was at least half a degree greater than it is at present. According to modern observations, the decrease at present is about 46" per century.

One degree of meridian for latitude 66° is stated to be about 365,800 English feet. Half a degree, therefore, is 182,900 English feet, equal to two million one hundred and ninety-four thousand eight hundred English inches (2,194,800).

To be able to state *that not one inch out of these two million and more inches,* which it is known the arctic circle has "shifted" during even so short a period as eighteen hundred years, can be accounted for on the supposition of an equatorial ocean two thousand feet in depth, and extending 45° each side of the equator, being turned into ice, and placed on the polar regions, is a valuable contribution to exact science.

This "*proof*" is valuable in more ways than one. It proves that our old earth is utterly independent of, and uninfluenced by, those dynamic laws which influence every other rapidly rotating body.

If a gyroscope top be spun, and we find that the axis of rapid rotation changes its direction in a certain way, we shall find that this direction changes almost immediately in another way when we alter the position of the centre of gravity by means of the smallest weight attached to the surface of the top. If, however, we could cause this top to rotate rapidly during one hundred years, a thousandth part of the weight would, during a long period, produce somewhat similar results to those produced by a heavier weight almost instantly.

It is, therefore, a very important item which is referred to by Dr. Croll, viz. that it is *proved* that, if the position of the centre of gravity of the earth be altered several thousand feet, yet the direction of the axis of rotation would "not" shift " by so much as an inch."

This is the theory which is put forward as a proof of—what? Probably that the obliquity can vary only between very narrow limits; but what are these limits? and by what laws is this narrowness limited?

Without following these theoretical conclusions further, or even hinting that they are anything but quite correct, an important suggestion is at once presented to us.

It has been stated how, on certain assumptions, and on the theories based thereon, it is impossible that the earth ever did or ever can move in some particular manner. Would it not be a proceeding of more practical value if these theorists were to rigidly define how the earth does move and has moved even during the past 180 years. The poles of the axis of daily rotation have, during this period, traced arcs of about 1°; the north pole having traced an arc of 1° nearly towards the first point of Aries. What is the length of the arcs which the zenith of Greenwich has traced owing to this movement? and what is the direction of these arcs as regards six hours, twelve hours, and eighteen hours right ascension during 180 years?

What are the lengths and directions of the arcs which the zenith of 62° north latitude has traced on the sphere of the heavens, owing to the same movement of the earth, during the past 180 years, for meridians of six, twelve and eighteen hours right ascension.

Considering the present distribution of land and water on the earth's surface, it would be a useful practical problem to demonstrate where the centre of gravity of the earth was at present situated, and whether it coincided to one inch with the centre of the sphere.

According to Dr. Croll, it has been proved that, if we took 45° on each side of the equator, amounting to 90° in all, and a depth of water of two thousand feet, and converted this water into ice, and piled it all within a few degrees of one or both poles, we should not alter the direction in which the earth's axis now moves by one inch.

Let us assume that we pile this mass of ice within 5° of the north pole. 90° piled into 5° would give eighteen times the depth of two thousand feet, equal to thirty-six thousand feet deep of ice, giving extra weight around the north pole, and withdrawing this weight from the equatorial regions. Would this condition produce no change in the position of the centre of gravity of the earth, and hence, as a mechanical law, causing the earth's axis of rapid rotation *to change its direction*, in a manner slightly different from that in which it now changes it?

If it be asserted that, no matter how you change the position of the centre of gravity of a rotating sphere, yet you can by this change produce no alteration in the direction in which the axis of this sphere points, I must at once claim that the laws of dynamics and actual facts prove the very opposite to be the truth.

But why do certain learned theorists occupy their time in attempting to prove what *cannot* occur? Why do they not state what *does* occur and has occurred during the past two thousand years of which we have records?

The learned Sizzi *proved* to his own satisfaction, and also to that of the school of science to which he belonged, that it was impossible that Jupiter could possess satellites. Would it not have been a more practical proceeding if he had looked through a telescope and seen them?

Would it not be a more practical proceeding if theorists were to look at the second rotation of the earth, and prove for themselves that a knowledge of this movement enabled them to calculate for any length of time the polar distance of a star from one observation only? Surely such an investigation would be more useful than inventing vague theories, in order to endeavour to prove that something could not occur?

Would it not be a more practical proceeding *to define*

accurately every detail connected with the movement of the earth, which accompanies the change in position of the zenith of the pole, instead of leaving this movement as it is at present, vague and undefined, and inventing theories to account for something, when it has never been defined what this something really is?

With regard to this movement of the earth, we have to deal with facts, not with vague theories; we have to define how *each* portion of the earth moves during each year, not how the poles only of the earth move.

Certain gentlemen who have distinguished themselves by rushing in where more prudent persons would hesitate to tread, have stated that the exact geometrical investigation to which I have given many years of research is " an absurd theory."

I have examined carefully the various books and articles which I have written on this movement of the earth, and also the notes and reports of various lectures that I have given on the same subject. I can find no theory put forward by me in any one of these works.

That which I have done is to examine in detail the movement of a globe, which accompanies a change in direction of the axis of rapid rotation. I define how each point on the surface of this globe moves annually during thousands of years. I then prove that, by a knowledge of this movement, the polar distance of a star can be calculated for distant dates with minute accuracy, without reference to more than the one observation which determines the polar distance and right ascension of this star for one particular date.

Where is the "absurd theory"? If this rigid geometrical proof can be termed a theory, then it is a theory that two sides of a plane triangle are greater than the third side, and it is also a theory that the three triangles of an equilateral triangle are equal to each other.

As another example of the unreasoning arguments put forward as regards this geometrical investigation being "some theory," the following may be given.

Suppose an observer to be located in 51° north latitude and provided with an instrument by which he could measure zenith distances. By aid of this instrument he would find that his zenith had traced on the sphere of the heavens an arc which was 9° 25′ 48″ in length during one hour of time.

A geometrician might now inform this observer that if he had been located on the equator, instead of at 51° latitude, his zenith would have traced on the sphere of the heavens an apparently straight line 15° in length during one hour. Also if he had been located in 70° latitude, his zenith would have traced an arc 5° 7′ 48″ in length during one hour. A geometrician would know these facts in consequence of being acquainted with the *daily* rotation of the earth.

What opinion should we form of the intellectual or scientific qualifications of a writer who stated that this assertion of the zeniths of different latitudes tracing arcs of very different values was an "absurd theory" which is not worth considering?

A knowledge of the second rotation of the earth enables a geometrician to make similar exact calculations in connection with the second rotation that can now be made in connection with the daily rotation.

It can be stated that whilst the zenith of the locality in south latitude 29° 25′ 47″, and on a meridian of eighteen hours right ascension, is carried annually by the second rotation over an arc of 40·9″ in the exactly opposite direction to that in which the daily rotation carries this zenith, yet the zenith of a locality on the equator is carried annually by the second rotation for the same meridian over an arc of only 35·62″, in opposition to the daily rotation. The zenith of a locality in latitude 51° 28′ north, such as Greenwich, is carried annually by the second rotation for the same meridian of right ascension over an arc of only 6·47″. The zenith of Edinburgh is carried over an arc of 3·3″ only, whilst that of Petersburg, in north latitude 59° 56′, is carried over an arc of only 0·65″ annually by the second rotation, these arcs in each case being for eighteen hours right ascension in direct opposition to the daily rotation.

The daily rotation does not readjust these zeniths exactly in the relative positions which they before occupied, the daily rotation being round the pole of the heavens as a centre; the second rotation being round the pole of second rotation as a centre, this pole of second rotation being 29° 25′ 47″ from the pole of daily rotation.

A point on the equator is, for eighteen hours right ascension, carried over an arc of 35·62″ in opposition to the daily rotation; consequently, when 35·62″ of the *daily* rotation has occurred, this point occupies, as regards its zenith, exactly the same position it occupied previous to its displacement by the second rotation. This 35·62″ of daily rotation for the equator produces for the zenith of Greenwich a *daily* rotation represented by an arc of 22·17″.

But the zenith of Greenwich has been displaced by the second rotation by an arc of 6·47″ only during the year, whilst that of Petersburg has been displaced only 0·65″.

Will those gentlemen who have thought it desirable to term such investigations as the above "absurd theories," kindly inform those who may be desirous of knowing the actual facts connected with the earth's movements, how much the lengths of localities on the equator, in south latitude 29° 25′ 47″, in north latitude 51° 28′, and in north latitude 60° are displaced annually as regards a meridian of eighteen hours right ascension by their theory of "a conical movement of the earth's axis," which movement makes "a shift" "of this axis?"

It is a mere waste of time to discuss scientific questions with individuals who ignore even the elementary laws of geometry and dynamics, and who claim that this and the other has been *proved* by certain persons whom they put forward as "authorities."

In a work published in 1875, termed, "Climate and Time," the author, Dr. James Croll, makes the following statement :—

"The polar regions owe their cold, not to the obliquity of the ecliptic, but to their distance from the equator. Indeed, were it not for the obliquity, those regions would be much colder than they really are, and an increase of obliquity, instead of increasing their cold, would really make them warmer. . . . It therefore follows that, although the arctic circle were brought down to the latitude of London, so that the British Islands would become a part of the arctic regions, the mean temperature of these islands would not be lowered, but the reverse. The winters would no doubt be colder than they are at present, but the cold of winter would be far more than compensated by the heat of summer."

Having had considerable experience of the climates of various countries on the earth's surface, I believe that the meridian altitude of the sun and the length of time it remains above the horizon is the great cause of the various temperatures in different localities, and that the equator and the distance of localities from the equator is an effect, not a cause.

The annual variations of climate in Quebec and many other parts of Canada may be taken as an example. The latitude of Quebec is about 46° 48' north. In winter I have seen the thermometer at 30° F. below zero. I have seen freshwater lakes during twenty-four hours frozen ten inches thick with ice. If this locality were deprived of 12° more of the sun's meridian altitude, it would be equivalent to placing the locality 12° farther north, and giving to the locality a temperature some 15° or 20° more of cold, when, consequently, more ice and more snow would accumulate during winter. In summer, at Quebec, the thermometer rises to 90° F. in the shade; the ice and snow are rapidly melted, and the rush of water, bearing icebergs, etc., is very great. Increase the cold during winter by 15° or 20°, and the heat in summer by even a greater increase, and what would be the result? A vast accumulation of ice and snow rapidly melted and scattered over the country.

I fail, however, to perceive why, because Dr. Croll imagines that an enormous increase in the eccentricity of the earth's orbit occurred, he should assert that a rigid investigation of the actual movements of the earth is an "absurdity." I can assure this gentleman that, if he will examine the geometrical laws connected with the second rotation of the earth, he will find that the assumed variation is even, in its details, based on a geometrical error.

But what have such theories to do with the actual movements of the earth? State how the earth does move and has moved during the past two thousand years even, is a fair question. State over what length of arc and in what direction the zeniths of location 10°, 30°, 50°, and 90°, from the north pole have been carried during even the past 180 years by "a conical movement of the earth's axis" as regards six, twelve, and eighteen hours right ascension.

Because a gentleman invents a theory to account for some effects, which theory is based on a geometrical error, why should he state that to rigidly examine and define those movements which have never hitherto been defined is an "absurdity"? When this and any other theorist can calculate the polar distance of a star for a hundred years or more from one observation only, then they may claim to have some knowledge of the earth's movements. As, however, such a calculation is unknown to them, it really appears that they are expressing their opinions on a problem which they have not examined or even understood.

It is no part of the work of a geometrician to join in the disputes of geological theorists. There are certain gentlemen who assert that, during the last glacial period, the earth from middle latitudes to the poles was covered by a cap of ice several thousand feet in thickness which was never exposed to any great heat. Theories have been invented to account for this imaginary cap of ice.

As stated in my late work, "Thirty Thousand Years of the Earth's Past History," both Sir John Lubbock and Professor Tyndall state that heat is as necessary as is cold to produce glaciers. Consequently, this cap of ice is a mere theory, on which learned gentlemen differ. It really appears difficult to account for vast masses of floating ice being carried over the land unless by the aid of heat. To prove what the movements of the earth really are and have been during many hundred years is the work of a geometrician. If these movements do not agree with the speculations of theorists, it is so much the worse for the theories.

Certain theorists have stated that, if the obliquity of the ecliptic were so great as to cause the arctic circle to reach, as it does in Venus, to within 15° of the equator, the *mean* annual temperature of all localities in middle and high latitudes would be much greater than it is at present. In the first place, this is a mere assertion. We know so little as to the effects which would be produced by the melting of vast masses of ice and snow, that positive statements are mere assumptions. Whether *all* the ice formed each winter would be melted each summer is a question which no man can answer positively. Secondly, in the St. Lawrence river, under present conditions, I have seen masses of ice of twenty feet cube formed, not during one winter, but during one month of intense cold. Give several degrees more cold, which would result from 12° more of obliquity, and much larger icebergs would be formed.

The *mean* annual temperature for any locality is no criterion of the climate. There are several localities where the mean annual temperature is about 60° F., the heat in summer being about 70° F., that in winter about 50° F. When the obliquity was about 35°, there might be localities where the mean annual temperature was 60° F., the winter cold being some 70° below zero, the summer heat some 130° F. Water frozen at a temperature considerably below zero takes a long time to thaw, and masses of ice thus thawed would produce fogs, as they do on the banks of Newfoundland, and thus veil the sun, and prevent its full power from being felt.

The conclusions formed as to what must happen under certain conditions are mere theories, which may or may not be near the truth; thus, although reference is made to the supposed conditions which theorists assume must follow certain variations in the obliquity, it is necessary to remember that the statements put forward as facts are merely guesses.

At p. 417, "Climate and Time," the author makes the following remark: "But even supposing it could be shown that a change in the obliquity of the ecliptic, to the extent assumed by Mr. Belt and Lieut.-Colonel Drayson, would produce a glacial epoch, still the assumption of such a change is one which physical astronomy will not permit."

This hitherto is quite correct. Physical astronomy and

its theories "will not permit" any change in the obliquity beyond very narrow limits, and for the following reasons: Physical astronomy does not deal with geometry. Physical astronomy has never yet defined how the earth has and does move. It has never even referred to the detail movements of various zeniths and meridians which accompany the change in direction of the earth's axis. Physical astronomy has been content to accept "a conical movement of the earth's axis" as a full and complete definition of all the detail movements which are produced by the second rotation.

Physical astronomy cannot calculate the position of any star even for one year from one observation of this star. It claims to be able to state what is the density of the various planets, such as Mercury, Venus, Mars, Jupiter, etc. The proof as to the accuracy or otherwise of these speculations does not exist. But physical astronomy omits to give any explanation as to why the axes of daily rotation of these planets are inclined at such very different angles to the planes of their orbits. It makes no mention of what sort of conical movement these axes make, or whether they make any movement.

Let us now refer to something more practical, and capable of being tested as to its accuracy.

In the Nautical Almanac for 1887 the mean right ascension and the south declination for January 1, 1887, of the star β Corvi are recorded as found by observation at that date as follows:—

Mean right ascension.	Mean south declination.
12h. 28m. 26·99s.	22° 46' 17·85"

Let physical astronomy, or any branch of astronomy hitherto known, calculate the mean declination of this star for January 1, 1850, and January 1, 1780, without reference to any other observation of this star.

It is almost needless to state that astronomers hitherto have not been competent to make such a calculation.

When the second rotation of the earth is understood, the problem is easily solved, and the solution is as follows:—

From the above observation we obtain the distance of this star from C, the pole of second rotation = 106° 19′ 15·2″. We know that the distance between the pole of second rotation and the pole of daily rotation is 29° 25′ 47″.

From the one observation given above, we can calculate the angle at C for January 1, 1887, this angle being 107° 33′ 52″. This angle is increased annually by the second rotation at the rate of 40·9″. At the date 1780 the angle, consequently, was less than it was in 1887 by 40·9″ × 107 = 1° 12′ 56″. The angle at C for January 1, 1780, was 106° 20′ 56″. With the two sides, viz. C P = 29° 25′ 47″, C β = 106° 19′ 15·2″, and the included angle P C β = P C β = 106° 20′ 56″. The third side P β, the polar distance of the star β Corvi, can be calculated, and will be found to be 112° 10′ 38″; from which take 90°, the distance of the equator from the pole, and we obtain 22° 10′ 38″ for the south declination of β Corvi for January 1, 1780.

In a catalogue of stars now before me, the south declination of this star for January 1, 1780, is recorded as 22° 10′ 38″.

This is not the kind of test that physical astronomy cares to deal with, nor has it hitherto defined the detail movements of the earth. The assumption that physical astronomy cannot "permit" this or that, is a similar claim to that of the theorists who would not "permit" Jupiter to possess satellites, nor the earth to have a daily rotation.

Further discussions on such classes of arguments and objections are unnecessary. It is a mere waste of time to have to point out the errors of pompous assertions, and to prove the feebleness of objections urged against the second rotation of the earth, and the facts proved thereby, when, as is the case, exactly similar objections were brought against the daily rotation in former times. The reader will find it a more useful occupation to examine the details of the second rotation, and to prove for himself that by a knowledge of this movement he can solve problems hitherto entirely beyond the powers of present theorists, who claim infallibility, than to devote his time and attention to replying to objections which have no foundation other than the beliefs and assertions of certain persons put forward as "authorities."

It has been truly remarked by Seneca, that men would rather hold fast to an error they know to be false, in order to keep up the appearance of not having been deceived, than avow they were in the wrong.

Thus we find writers asserting that "*a* conical movement of the earth's axis round the pole of the ecliptic as a centre" is a statement accurate and detailed of the movements of the earth. This assertion is accepted as complete, although it has never been stated which pole remains fixed, how each zenith and meridian is affected, in what direction the centre of gravity of the earth is thrown out of its orbit by the conical movement of *the axis*, etc.

It may, however, be of interest to the geometricians and astronomers of the future if the so-called arguments and proofs of objections against the second rotation of the earth be collected and preserved as specimens of the scientific reasoning of the present day. We possess examples of the arguments used in former times against the daily rotation of the earth, and against the possibility of Jupiter possessing satellites; it will be curious to compare the ancient with the modern objections, and to discover, if possible, where they differ.

www.ingramcontent.com/pod-product-compliance
Lightning Source LLC
Chambersburg PA
CBHW031903220426
43663CB00006B/743